影印版说明

《传感材料与传感技术丛书》中第一个影印系列 MOMENTUM PRESS 的 *Chemical Sensors: Fundamentals of Sensing Materials & Comprehensive Sensor Technologies*（6卷，影印为10册）2013年出版后，受到了专家学者的一致好评。

为了满足广大读者进一步的教学和科研需要，本次影印其 *Automotive Sensors*。

材料科学与工程图书工作室

联系电话 0451-86412421
　　　　 0451-86414559

邮　　箱 yh_bj@aliyun.com
　　　　 xuyaying81823@gmail.com
　　　　 zhxh6414559@aliyun.com

AUTOMOTIVE SENSORS

EDITED BY **JOHN TURNER**

影印版
汽车传感器

哈尔滨工业大学出版社
HARBIN INSTITUTE OF TECHNOLOGY PRESS

黑版贸审字08-2014-076号

John Turner
Automotive Sensors
9781606500095
Copyright © 2009 by Momentum Press, LLC
All rights reserved.
Originally published by Momentum Press, LLC

English reprint rights arranged with Momentum Press, LLC through McGraw-Hill Education（Asia）

This edition is authorized for sale in the People's Republic of China only, excluding Hong Kong, Macao SAR and Taiwan.

本书封面贴有McGraw-Hill Education公司防伪标签，无标签者不得销售。
版权所有，侵权必究。

图书在版编目（CIP）数据

　　汽车传感器＝Automotive Sensors：英文/（英）特纳（Turner, J.）主编. —影印本. —哈尔滨：哈尔滨工业大学出版社，2015.1
　　（传感材料与传感技术丛书）
　　ISBN 978-7-5603-4908-4

　　Ⅰ.①汽… Ⅱ.①特… Ⅲ.①汽车传感器-研究-英文 Ⅳ.①U463.6

　　中国版本图书馆CIP数据核字（2014）第242659号

责任编辑　杨　桦　张秀华　许雅莹
出版发行　哈尔滨工业大学出版社
社　　址　哈尔滨市南岗区复华四道街10号　邮编150006
传　　真　0451-86414749
网　　址　http://hitpress.hit.edu.cn
印　　刷　哈尔滨市石桥印务有限公司
开　　本　787mm×960mm　1/16　印张18
版　　次　2015年1月第1版　2015年1月第1次印刷
书　　号　ISBN 978-7-5603-4908-4
定　　价　90.00元

（如因印刷质量问题影响阅读，我社负责调换）

CONTENTS

1	Sensors in the Vehicle *John Turner*	1
2	Automotive Pressure Sensors *M. J. Tudor and S. P. Beeby*	37
3	Temperature Sensors *John Turner*	85
4	Automotive Airflow Sensors *John Turner*	107
5	Combustion Sensors *Peter Eastwood*	115
6	Automotive Torque Sensors *John Turner*	159
7	Displacement and Position Sensors *Jonathan Swingler*	175
8	Accelerometers *Jonathan Swingler*	203
9	Gas Composition Sensors *Jonathan Swingler*	231
10	Liquid Level Sensors *Yingjie Lin and Francisco J. Sanchez*	259

PREFACE TO THE
SENSORS TECHNOLOGY SERIES

The creation of the digital world is perhaps the most remarkable engineering event of the late twentieth century and may be compared in impact to the harnessing of steam at the beginning of the nineteenth. Obviously, the two cannot otherwise be compared, representing as they do the transition from the era of power to that of information. Even the term "transition" is misleading because the digital world depends entirely on the prior existence of sources of electrical energy, currently derived largely from thermal and nuclear processes, with an ever-increasing reliance on renewable content including the harnessing of tidal, wind and solar sources. Furthermore, chemical sources of energy should not be forgotten, including the wide range of "personal" batteries powering everything from laptop computers to hearing aids and pacemakers, and the currently-developing fuel-cell technology that may yet make the electric automobile a viable proposition.

The digital world is also dependent upon the provision of sources of information. Such input information can itself be digital, as when a human being types data directly into a computer or when nuclear disintegrations are counted by an appropriate sensor. However, other sources of information rely upon sensors that transduce phenomena derived from mechanical, optical or chemical/biological phenomena into (usually) electrical signals. Because real-world phenomena are predominantly analog in nature, these signals are therefore also in analog form. Such signals may then be amplified, conditioned and applied to analog-to-digital (A-D) converters for entry into the digital world—all major topics in their own right. Actually, there is always a level at which even digital processes take on analog characteristics. For example, the heat generated within computer chips arises mainly during the (analog) transitions between the ON and OFF states of the multitude of transistor structures contained therein.

The present series of volumes is entirely concerned with sensors themselves, and because the subject matter is so wide-ranging in both scope and maturity, this is reflected within the individual books. So, whereas care has been taken to include a considerable amount of practical material, the proportion of such leavening is inevitably variable. Thus, the volume concerning chemical sensors may be regarded as largely oriented towards research and development, whereas at the other end of the spectrum, the avionics volume and the present automotive volume are based largely on current practice. This disparity is in part driven by the safety-oriented conservatism in these latter fields, but future developments in both have not been ignored.

Though the gestation period of such a comprehensive series has been long, care has been taken to include information that is indeed both basic and also contemporary, so providing a platform for continued updating as progress continues.

The ever-increasing sensor/electronics content of a modern automobile is reflected in the proportion of the ultimate cost of the vehicle related to this content, and this is also true for equally-modern aircraft. In both cases the sensors currently employed in production models are well-tried, safe and reliable, and so command the bulk of the material presented here. The intention of both volumes is to provide the reader with an authoritative introduction to the theory, installation, and performance of these sensors prior to discussing future developments. An excessive use of theoretical or mathematical material has been avoided throughout, and it is hoped that this will lead to a unique "hands-on" approach.

The present volume begins with an overview of the sensor/electronic content of the modern automobile by the Volume Editor, Prof. John Turner, and each of the topics mentioned is then covered in greater detail in the succeeding chapters. All the authors are practicing automobile engineers currently working in industry or academia, and each chapter therefore represents an authoritative viewpoint gleaned from personal experience.

J. Watson
Editor-in-Chief
February 2009

CHAPTER 1

SENSORS IN THE VEHICLE

John Turner

1. INTRODUCTION

Modern vehicles can have up to one hundred microcontroller-based electronic control units (ECUs). These are fitted to enhance safety, performance, and convenience. As the complexity and safety-critical nature of automotive control systems increases, more and more sensors are needed.

The problems of congestion and environmental effects associated with increasing vehicle ownership and use are well known. Throughout the world, the climate of opinion is turning against unfettered mobility. In most developed countries, government transport policy places increasing emphasis on the efficient management of existing roads and recognizes the difficulty of satisfying demand by building new roads. The use of a combination of information technology (IT) and electronic systems to advise drivers and reduce their workload is being seen as offering at least a partial solution to current and projected traffic problems, since it has been shown to smooth the traffic flow [1]. Advanced sensing and control systems also offer the possibility of safely reducing the gaps between vehicles (the "headway"), thus increasing the number of vehicles that can use a particular stretch of road.

The combination of IT and vehicle-based and highway-based electronic systems has become known as *automotive telematics*. All telematic systems rely heavily on sensors and measurement techniques, and this is especially true of those applications that are safety critical. Many research projects are currently underway in this area, which is particularly challenging for the sensor designer in the light of typical automotive cost constraints. (As a rule of thumb, at 2005 prices, a vehicle manufacturer will normally tolerate a "measurement cost" of only around $10 per measurand, including all the signal conditioning required). If an automotive sensor costs significantly more than this, the extra cost has to be justified in terms of additional functionality, perhaps because the measurement can be used for several purposes.

Highway sensor costs are much higher than those of automotive devices. There are two main reasons for this. First, the volumes are much lower—many highway sensors are almost custom made for a particular application, and volume production runs of any particular configuration are rare. Second, the cost of the associated groundwork has to be taken into consideration. These factors mean that the cost of installing, for example, a loop detector (explained in section 3.1) is typically several thousand dollars.

To succeed commercially, automotive sensors have to be very robust. They must tolerate an environment that includes temperatures from −40 to +140 °C, possible exposure to boiling water, battery acid, fuel, hydraulic fluid, road salt, and so forth, as well as very high shock and vibration loads, which can exceed 1000 g on the unsprung side of the vehicle suspension. They may also have to tolerate and function in the presence of high levels of electromagnetic noise.

Highway sensors also have to be robust. They too experience the full range of climate conditions, and may also occasionally be exposed to fluids originating from motor vehicles. In general, however, the shock and vibration environment is less demanding, and the cost constraints are usually less extreme.

2. ON-VEHICLE SENSORS

A good example of the trend towards increasing complexity is provided by the air bag system. Early (1980s) systems typically used an accelerometer and a "safing sensor" (to avoid inappropriate firing of the airbag, for example, when the vehicle is stationary). Current designs may include sensors for child seat detection, seat position, occupant position, occupant detection, and vehicle speed. Rollover sensors, side impact sensors, weight sensors, tire inflation sensors, and tire temperature sensors may also be used. Several sensors may be required to monitor different seats or zones within the vehicle.

A similar increase in sensing complexity has taken place in powertrain control systems, vehicle body controllers (e.g., door locks, windows, sunroofs, and wipers), and chassis control systems. This has produced significant benefits in system performance, but has also created some real challenges.

A complex electronic/electromechanical system, such as a modern motor vehicle, has to be operated in intimate varying interaction with its driver and with an outside world of considerable complexity. For any such system to operate satisfactorily, the need for effective, accurate, reliable, and low-cost sensors is very great. Electronic measurement systems can be applied very widely within a motor vehicle, as shown in figure 1.1. The complexity can range from the interactive control of engine and transmission to optimize economy, emissions, and performance to the simple sensing of water temperature and fuel level.

2.1. POWERTRAIN SENSORS

The complexity of the control task involved in powertrain management is demonstrated by figure 1.2. Table 1.1 lists typical required specifications for the powertrain sensors. The accuracy and temperature range over which these devices have to operate should be noted, and it also must be remembered that they have to be of minimal cost and high reliability. A typical automotive sensor has a design life of up to 10 years, and should require no initial setting up or maintenance within that time. A fully comprehensive powertrain control system would contain most of the devices listed in table 1.1, although the really critical devices are those which measure engine timing, inlet manifold mass airflow, manifold vacuum pressure, exhaust gas oxygen content, transmission control valve position, transmission input and output speed, and throttle and accelerator position.

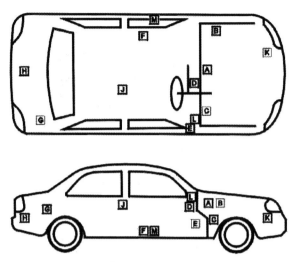

Figure 1.1. Automotive measurement systems

2.1.1. Ignition Control

The ignition timing sensors available at present normally use Hall effect [2] or other electromagnetic transducers to detect the movement of a magnet or metallic projection attached to the flywheel. The major inaccuracies in ignition timing arise from mechanical vibration and torsion in the geared drive to the distributor from the engine crankshaft. This problem is likely to be partially overcome in future vehicles, where the timing is taken directly from the crankshaft, although this then requires an additional sensor on the camshaft to determine the correct timing for each cylinder in the four-stroke cycle. Crankshaft sensing itself may suffer from windup errors due to the main engine torque and the influence of differing operating conditions in each cylinder.

The measurement of inlet manifold vacuum pressure was the first sensing requirement in early ignition control systems [3], and it continues to be a very important parameter. It provides a relatively good measurement of engine torque, since as the engine slows down under load the inlet manifold vacuum pressure moves closer to atmospheric pressure. This effect is accentuated by the driver (who is part of the control loop) pressing the accelerator and opening the throttle further.

It would be much better to measure the engine torque directly, if a reliable low-cost way to do this could be found. The search for a low-cost, noncontact method of torque measurement is currently the subject of a great deal of automotive research, as discussed in chapter 6.

The inferred measurement of load by sensing manifold vacuum has been used for controlling ignition advance from a very early stage in the development of the internal combustion engine. For many years the preferred approach was mechanical. The load-related control was achieved through the use of an aneroid vacuum capsule connected to the manifold. The varying vacuum altered the aneroid capsule shape, producing a force that physically rotated the distributor to alter the ignition advance angle. At the same time, a centrifugal weight system further controlled ignition advance angle according to the rotational speed of the engine.

In an ignition system with electronic control, these functions are taken over by a pressure transducer connected to the manifold, and a measurement of the engine rotational speed is obtained from a

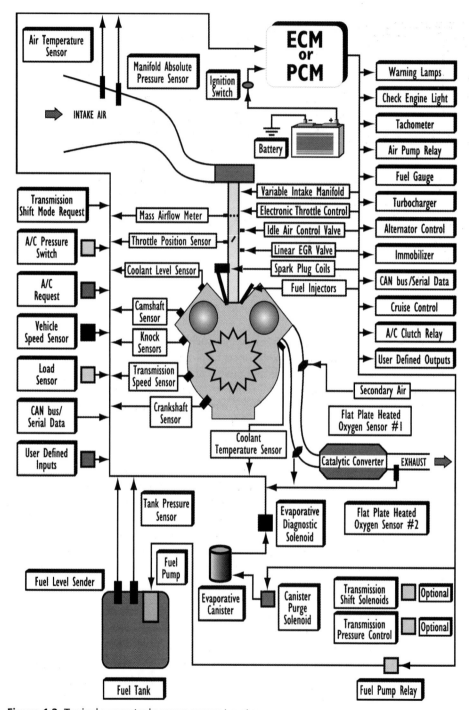

Figure 1.2. Typical powertrain management system

Table 1.1. List of powertrain sensor specifications, from reference 2

Sensor/type	Sensing method	Range	Accuracy (%)	Thermal range (°C)	Response time
Inlet manifold absolute or differential pressure sensor (gas engines)	Piezoresistive silicon strain gauged diaphragm *or* capacitive silicon diaphragm	0–105 kPa	±1 at 25 C	−40 to +125	1 ms
Inlet and exhaust manifold pressure sensor (diesel engines)	As above	20–200 kPa	±3	As above	10 ms
Barometric absolute pressure sensor	As above	50–105 kPa	±3	As above	10 ms
Transmission oil pressure sensor	Differential transformer and diaphragm, *or* capacitive diaphragm (often stainless steel)	0–2000 kPa	±1	−40 to +160	10 ms
Inlet manifold air temperature sensor	Metal film or semiconductor film	−40 °C to +150 °C	±2 to ±5	−40 to +150	20 ms
Coolant temperature sensor	Thermistor	−40 °C to +200 °C	±2	As above	10 s
Diesel fuel temperature sensor	Thermistor	−40 °C to +200 °C	±2	40 to +200	10 s
Diesel exhaust temperature sensor	Cr/Al thermocouple	−40 °C to +750 °C	±2	−40 to +750	10 s

TABLE 1.1 CONTINUED

Ambient air temperature sensor	Thermistor	−40 °C to +100 °C	±2	−40 to +100	10 s
Distributor-mounted timing/speed/trigger sensors	Hall effect *or* optical *or* eddy current *or* variable reluctance	Zero to maximum engine speed	±1	−40 to +180	N/A
Crankshaft-mounted timing/speed/trigger sensors	As above	As above	±1	−40 to +180	N/A
Road speed sensor: −speedometer cable/ gearbox fitting	Optical *or* Hall effect *or* reed switch	As above	±1	−40 to +125	N/A
Speed-over-ground sensor	Optical *or* radar	Zero to max. vehicle speed	±1	−40 to +125	N/A
Inlet manifold air-mass flow (unidirectional)	Vanemeter *or* hot wire *or* hot film	10 to 100 kg/h *and* 20 to 400 kg/h (two ranges)	±2 ±2	−40 to +125	35 ms for vanemeter, 5 ms for others.
Inlet manifold air-mass flow (bidirectional)	Ultrasonic *or* corona discharge *or* ion flow	±200 kg/h	±2	−40 to +125	1 ms

TABLE 1.1 *CONTINUED*

Accelerator pedal position sensor	Potentiometer	0–5 k from min. to max. Pedal travel	±1	−40 to +125	N/A
Throttle position sensor	Potentiometer	0–4 k from closed to open throttle	±3	−40 to +125	N/A
Gear selector position sensor	Microswitch *or* potentiometer	8 position selection *or* 0–5 k	±1	−40 to +150	N/A
Gear selector hydraulic valve position sensor	Optical encoder	8 position selection	±2	−40 to +125	N/A
EGR[1]* valve position sensor	Linear displacement potentiometer	0–10 mm	±2	−40 to +125	N/A

TABLE 1.1 CONTINUED

Closed throttle/full throttle sensors	Microswitches	N/A	N/A	−40 to +125	N/A
Engine knock sensor (gas engines)	Piezoelectric accelerometer	5–10 kHz, up to 1000 g	N/A	−40 to +125	Depends on resonant frequency
Engine knock and misfire sensor	Ionization measurement in cylinder or exhaust manifold	N/A	N/A	−40 to +150 (externally). Probe exposed to combustion gas temperatures	20 μs
Exhaust gas oxygen content sensor	Zirconium dioxide ceramic with platinum electrodes *or* titanium disks in aluminum	50% to 150% stoichiometric air-fuel ratio	3%	+300 to +850	50 ms
Exhaust gas oxygen content sensor for lean-burn operation	Zirconium dioxide oxygen pumping device with heater	14:1 to 30:1 air-fuel ratio	1%	+300 to +850	50 ms

TABLE 1.1 CONTINUED

Closed throttle/full throttle sensors	Microswitches	N/A	N/A	−40 to +125	N/A
Engine knock sensor (gas engines)	Piezoelectric accelerometer	5–10 kHz, up to 1000 g	N/A	−40 to +125	Depends on resonant frequency
Engine knock and misfire sensor	Ionization measurement in cylinder or exhaust manifold	N/A	N/A	−40 to +150 (externally). Probe exposed to combustion gas temperatures	20 µs
Exhaust gas oxygen content sensor	Zirconium dioxide ceramic with platinum electrodes *or* titanium disks in aluminum	50% to 150% stoichiometric air-fuel ratio	3%	+300 to +850	50 ms
Exhaust gas oxygen content sensor for lean-burn operation	Zirconium dioxide oxygen pumping device with heater	14:1 to 30:1 air-fuel ratio	1%	+300 to +850	50 ms

* Exhaust Gas Recirculation

sensor connected to the crankshaft. The pressure and speed signals provide inputs to a microprocessor, which is programmed to look up the optimum advance angle from a three-dimensional table relating speed, load, and advance angle stored in the microprocessor's memory (see figure 1.3 for an example). By this means significant improvements in engine operation and economy can be obtained. A number of designs of manifold pressure sensor have been used for this system, including devices based on capacitive, inductive, and potentiometric techniques. The most widely used approach is to employ a silicon diaphragm with integral silicon strain gauges, or to use a capacitive deflection sensing method [4]. In both of these sensors, a disk of silicon is etched to form a thin diaphragm (see figure 1.4) to which the pressure is applied. The strain gauges are integrated onto the disk, or a second capacitive plate is added. This technique produces a reliable low-cost device with a good resistance to the high-temperature, high-vibration conditions under which it has to operate.

2.1.2. Knock Sensing

When an ignition system with electronic advance control is optimized for best performance and economy, it can, under some conditions, be set sufficiently far advanced to cause a condition known as "knocking." Under these conditions premature high-rate combustion ("detonation") takes place, which, because of the rapid pressure increase, can quickly cause physical damage to vulnerable structures within

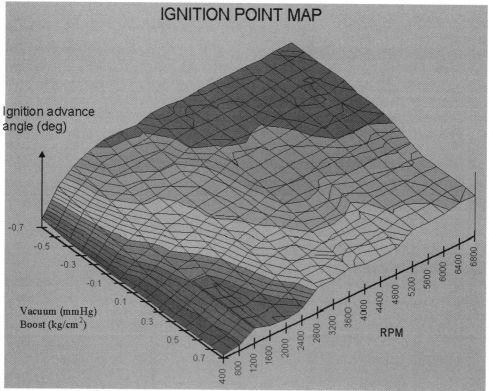

Figure 1.3. Ignition timing map

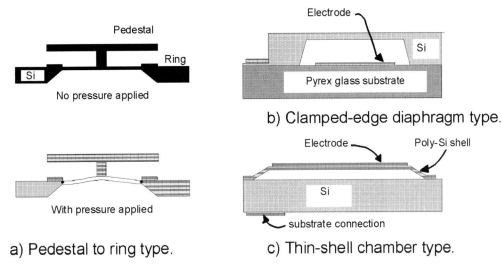

Figure 1.4. Silicon capacitive pressure transducer structures used in automotive engineering

the combustion chamber, such as the piston crown. In an engine with conventional mechanical ignition timing control, the advance angle is normally retarded sufficiently to avoid this condition, but considerable loss of efficiency results. For this reason, it is desirable to operate an electronically controlled ignition as close to the knock limit as possible, while retaining the ability to retard the ignition within one or two engine cycles to a safe level. This requires a method of rapidly sensing the knock condition, and to date this has usually been achieved by the use of a piezoelectric accelerometer, known, not surprisingly, as a knock sensor (see figure 1.5). An automotive knock sensor is usually mechanically tuned to be sensitive to the characteristic knock ringing frequency, which in normal-sized engines is in the region of 8 kHz. The transducer is positioned on the engine block in a place shown by extensive vibration analysis to give the best knock signal from all cylinders.

It has been proposed that measuring the ionization current across the spark plug after normal firing could provide an alternative method of obtaining this knock signal [5]. This could be done by applying a small voltage to the plug, sufficient to maintain an ionization current across the plug electrodes. This current has been shown to exhibit a superimposed ringing signal during knock, which can be distinguished and used to control ignition retard. This oscillation in ionization current appears to be due to the variations in plasma density caused by the pressure resonance initiated by the knock condition.

2.1.3. Fuel Control

The principle of using a three-dimensional look-up table as a means of describing optimized engine operation has been taken further with the electronic control of engine fueling. Most modern vehicles use electromagnetic fuel injectors, where the major parameters are again load and speed, but with the amount of fuel required being determined by the injector opening time as the controlled parameter. In this system, both the total quantity of fuel and the ratio of air mass to fuel injected into the engine are critical. Under these circumstances, the ideal measurement to be made is the mass airflow into the engine manifold. Measuring inlet manifold vacuum pressure, and then calculating the swept volume of

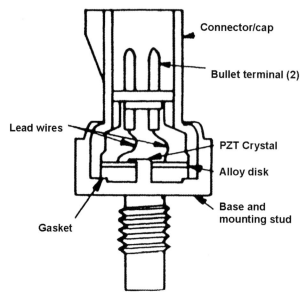

Figure 1.5. Antiknock sensor

an engine from its rotational speed, can give a reasonable estimate of mass airflow [6]. However, because of the compressibility of air and the significant volume of the manifold, this is a difficult calculation to make accurately, particularly under transient conditions. Many attempts have been made to develop a low-cost sensor to estimate air-mass flowrate directly, but it has always been a difficult measurement to make. The air vane-meter, in which a pivoted vane is placed in the airstream and attached to a single potentiometer as the measuring device (see figure 1.6), was the first sensor to be used in production and has been very widely applied. Its performance suffers from the fact that it measures air velocity rather than mass, and therefore the signal requires processing with a further signal representing air density to obtain air-mass flow. The transducer also has significant mechanical inertia under transient conditions, and there is a reduction of engine efficiency caused by the flap partially blocking the airflow into the manifold. However, the device has proved to be a very valuable sensor for a generation of fuel-injected engines and is still in production.

An alternative device now in common use is based on the hot-wire or hot-film anemometer (see figure 1.7). The mass flow of air reduces the temperature of an electrically heated wire or film, and, as a result, its resistance falls (metals have a positive temperature coefficient). This resistance change can be measured by, for example, a bridge circuit with appropriate amplification of the signal. See chapter 4 for further details.

The hot-wire or hot-film anemometer measures mass airflow directly, is fast in response (1–2 ms), and does not significantly obstruct the manifold. The sensor does, however, require a correction for ambient air temperature, and is also susceptible to contamination of the hot surface. This contamination is often dealt with by an automatic burn-off operation, which heats the wire to red heat on each occasion the vehicle is used. Hot-film transducers also suffer from inaccuracies under pulsed flow conditions, owing to their inability to differentiate the direction of flow.

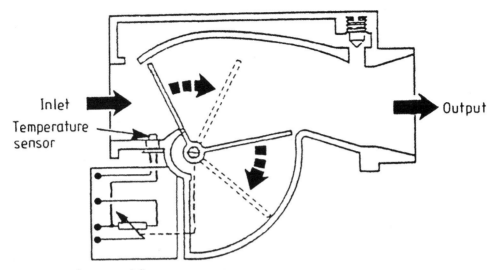

Figure 1.6. Moving-vane airflow sensor

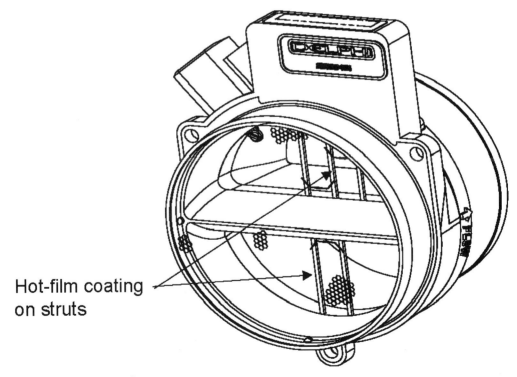

Figure 1.7. Hot-film airflow sensor

Several other sensors for the measurement of mass airflow into the manifold have been developed, notably the vortex shedding flow meter [7,8] and the ion drift flow meter [9]. This latter device has the major advantage of being able to measure direction of flow as well as its quantity. However, neither of these devices has proved entirely successful, although a vortex-shedding device has been used in at least one Japanese production vehicle. Further discussion of this appears in chapter 4.

2.1.4. Emission Control

In most developed countries, severe restrictions are placed on exhaust emissions. This makes the requirement to control the air-fuel ratio provided to the engine very critical. With current engine technology, the only way to meet the emissions regulations is to use a so-called three-way catalyst in the exhaust system to reduce the levels of the critical pollutants of carbon monoxide (CO), hydrocarbons (HC), and the oxides of nitrogen (NO_x). For its correct operation, such a catalyst requires that the air-fuel ratio supplied to the engine should always be as close as possible to the optimum stoichiometric level of 14.7:1. This is achievable only by the use of a feedback control system in which this air-fuel ratio is sensed by means of an exhaust gas oxygen sensor in the engine exhaust manifold.

This sensor (see figure 1.8) makes use of the fact that the migration of oxygen ions across a suitable membrane or filter from one gas to another is only dependent on the partial pressure of oxygen in the two gases. At the stoichiometric air-fuel ratio, the partial pressure of oxygen in the engine exhaust gas equals that in ambient air. If suitable electrodes are attached to each side of the ceramic filter used in the sensor, a positive or negative voltage is generated by ion migration when the air-fuel ratio is below or above stoichiometry. A rapid voltage change occurs over the transition between those two conditions. This voltage transition is ideal for use as a feedback signal to control the amount of fuel injected and, therefore, the air-fuel ratio. The sensor is known as an exhaust-gas oxygen (EGO) sensor or Lambda sensor (from the Greek letter Lambda conventionally used to denote stoichiometry).

Unfortunately, most exhaust gas oxygen sensors require measurements taken over a number of engine cycles in order to respond accurately, so sudden changes in engine speed or load may have to be compensated for by open-loop adjustments based on a three-dimensional map similar to that used for ignition timing. For stoichiometric control, such a map has to relate transient fueling to inlet manifold pressure, or mass airflow, and engine speed.

Until recently, future developments in emission-controlled engines, particularly in Europe, were expected to be towards the use of "lean burn" technology to obtain low levels of engine emissions, rather than by the use of the expensive three-way catalyst with a stoichiometric engine and with consequent poor fuel economy and performance. Lean-burn operation requires effective control of the engine at air-fuel ratios between 14:1 and 22:1 while still maintaining adequate drivability. If this is to be done by feedback methods, then the availability of a lean-burn exhaust gas oxygen (EGO) sensor becomes critical. Prototype sensors are available and use a technique known as "oxygen pumping" [10–12]. The oxygen pumping approach relies on the fact that applying a voltage across a filter forces oxygen ions to migrate to or from a small pumping cell, which positively controls the oxygen partial pressure within the sensor. A conventional EGO sensor is used to compare the increased partial pressure of oxygen in the cell with the increased partial pressure of oxygen in the exhaust gas stream from a lean-burn engine. The output voltage change at balance behaves like that from a conventional exhaust oxygen sensor. The current required to produce the balance condition in the pumping cell is a measure of the increase above stoichiometry of the air-fuel ratio being supplied to the engine.

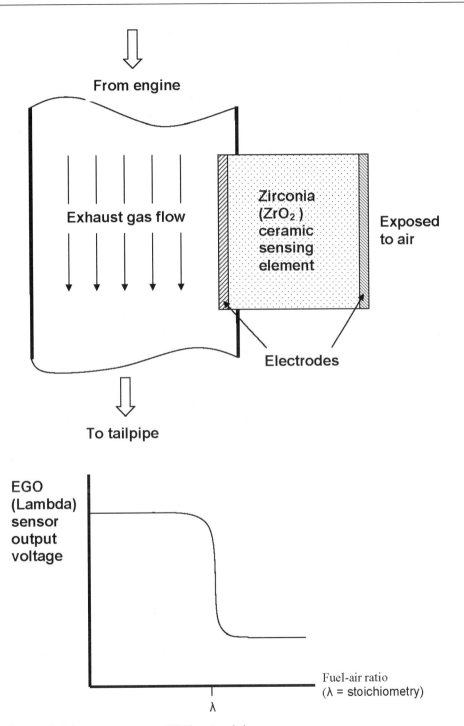

Figure 1.8. Exhaust gas oxygen (EGO) or Lambda sensor

A lean-burn oxygen sensor of this type still suffers from a response time that is long compared with the rate of change of engine conditions, and the associated control system therefore continues to require open-loop compensation for transient conditions.

The imposition of strict exhaust emission regulations in Europe has left lean-burn technology unable to meet the very low emissions levels now required. However, proposals have been made for mixed-cycle engines that run at stoichiometry when necessary to meet instantaneous emission levels, and under lean-burn conditions at other times (such as in high-speed driving). These mixed-cycle engines are reputed to give excellent fuel consumption and may bring back the need for a lean-burn oxygen sensor. Such developments could also increase the need for specific exhaust gas sensors for CO, HC, and NO_x levels to permit fine control of the switchover between stoichiometric and lean-burn operation.

2.1.5. In-cylinder Combustion Measurement

In an internal-combustion engine, a finely atomized fuel spray is dispersed in air, compressed within the cylinder, and ignited. The resulting fast burn of hot gases drives the piston down and is the source of the power produced by the engine. In a spark-ignition engine, the combustion is initiated by an electric arc from a spark plug. In a compression-ignition (or Diesel) engine, the heating effect of compression causes combustion. In both cases combustion starts at a point (in Diesel engines, often a "hotspot" on the piston crown), and the flame front spreads across the cylinder at the speed of sound.

In an ideal world, control of the process described above would be based on measurements of some meaningful property of the combustion process in the cylinder itself, undertaken quickly enough to allow the engine's operation to be controlled accurately on a cycle-to-cycle basis. Three methods have been developed for meeting this requirement. The first approach is based on measurements of the pressure variations within the combustion chamber. The second relies on detecting the arrival time of the flame front by means of an ionization detector, and the third technique assesses the optical output from the combustion process (i.e., its color and intensity) by means of an optical sensor.

High-cost, laboratory-grade piezoelectric pressure sensors have been available to the engine developer for many years. The problem in applying them to production vehicles has been the need to reduce the price to a level that will permit the economic fitting of a pressure sensor in each combustion chamber of the engine in mass production. The task is made more difficult by the extremely hostile environment within and close to the cylinder. This goal has now been approached by some sensor manufacturers, such as Texas Instruments [13], and a suitable piezoelectric sensor with a flush diaphragm has been developed for production use. Good correspondence has been shown between laboratory-quality high-cost instrumentation pressure sensors and these low-cost devices. (These sensors also offer the capability of detecting knock conditions in a combustion chamber; this is because knock is a ringing effect caused by a rapid pressure increase in the combustion chamber.)

In the case of an ionization sensor, the idea is to detect the arrival of the flame front of the burning air-fuel mixture on the far side of the combustion chamber from the point of ignition. This provides two pieces of information. First, the time it takes for the flame to arrive after the firing of the spark plug gives an indication of the suitability of the ignition timing, and can therefore be used as a feedback signal to correct that timing. Second, the scatter of arrival times between successive firings gives an indication of the air-fuel mixture weakness, because a weak mixture produces a greater variation in arrival times. This information can therefore be used as a measure of air-fuel ratio, and, when integrated over a number of engine cycles, can be used as a feedback signal to control the combustion process. These ionization measurements can also give an indication that knocking is taking place, as can measurements of the ionization across the spark-plug electrodes themselves after firing is complete.

An ionization sensor consists of a simple insulated electrode projecting into the combustion chamber in an appropriate location. A low voltage is applied between the electrode and the body of the engine. When the flame with its large supply of ions arrives, it reduces the resistance through the gas between the electrode and the surrounding metal, and generates a large and easily measured step change in current.

The third type of in-cylinder sensor is optical, and a number of studies have been done in this area. The approach most commonly adopted is to insert a quartz rod into the combustion chamber, which is viewed at the outer (cold) end by a photoelectric sensor. The parameters of interest are the variation in brightness and timing of the combustion. This data provides similar information, including some on knock, to that supplied by the pressure sensor.

2.1.6. Engine Speed and Torque Measurement

After the measurement of engine speed and timing, engine torque is probably the quantity of most interest to automotive engineers. The measurement of speed and timing has always been essential, since the control of even the very early engines required some events, such as ignition and valve opening, to take place at the correct time in the engine cycle. At present the distributor, which normally uses a self-generating electromagnetic pickup in the breakerless ignition systems found on most vehicles, provides engine speed data. With the advent of distributorless ignition systems, however, the use of digitizer disks on the main crankshaft and electromagnetic or optical sensors matched to these is expected to increase rapidly.

Low-cost methods of measuring engine torque have not so far become available for automotive use, and existing control systems have had to function without this important parameter. However, a number of new developments in low-cost torque measurement and its telemetry from moving to stationary parts of the vehicle have been made, and torque measurement may become important for the future of engine control. This will be essential if predictions on the use of embedded simulation come to fruition, since the accurate measurement of engine torque and speed and the comparison of these measurements with a good computer simulation should provide all the information necessary to control an engine effectively.

The engineer faced with the problem of measuring torque normally uses a torque transducer, which has been made by applying strain gauges to a shaft to measure the shear strain caused by torque. This type of transducer is widely used and probably forms the most common type of torque sensor. The major disadvantage of this approach for automotive applications is that additional equipment is usually required to transmit power to the rotating shaft and energize the strain gauge bridge, and also to retrieve the data. This apparatus can take the form of a set of slip rings, rotary transformers, or battery-powered radio telemetry equipment. Regardless of which is chosen, the need for some form of power and/or data transmission system makes the measurement of torque much more expensive than, say, that of pressure or temperature. In addition, slip rings (and to some extent rotary transformers) can be unreliable when operated in a dirty environment, and may be prone to radio-frequency interference (RFI). All of the above problems normally make strain gauge–based methods of torque sensing unsuitable for automotive applications. A number of other methods are potentially available, however, as described in chapter 6.

2.2. TRANSMISSION CONTROL

The function of the transmission in a road vehicle is purely that of a power-matching device between the power source (the engine) and the load. With a conventional manual transmission, the driver is part of the feedback loop, sensing speed and load and adjusting the transmission ratio within the mechanical limitations of the vehicle to what he or she perceives as being the best operating condition. One of the main feedback parameters used is engine speed, in the form of the pitch and noise level perceived by the driver. Unfortunately, this gives a rather poor representation of engine power output. Although the driver's gear changing may optimize subjective acceleration and drivability, it does not give anything like optimum operation for economy and performance. In fact, consideration of the torque-speed curves for a typical engine (see figure 6.3) shows that the optimum economy is obtained by keeping the engine at the lowest speed possible for as long as possible during acceleration, and changing the gear ratio to increase vehicle speed, only increasing engine speed (to produce more power output) when a wide-open throttle condition is reached. By this means the best possible economy is attained. Good acceleration performance will require some modification to this strategy.

Operating the transmission in this way requires the use of either an automatic stepped transmission with electronic control and smooth changes and sufficient steps to give an adequate range of gear ratios [14], or an electronically controlled, continuously variable transmission (CVT) with an adequately wide ratio.

The transmission ideally needs to be variable in ratio throughout the operating range, and, in a fully integrated powertrain (engine and transmission) control system, to be controlled interactively with the engine. The sensors required are for engine speed, transmission output speed, vehicle speed, and, assuming that the transmission ratio is hydraulically controlled, hydraulic valve position and hydraulic oil pressure. The first of these parameters is normally available from existing sensors on the engine. Vehicle speed is often available from an inductive sensor used on the transmission to provide a signal for an electronic speedometer. If a stepping motor actuates the hydraulic control valve, a separate sensor may not be required. If a sensor is required, an encoder disk and electromagnetic or optical sensor is attached to the end of the valve shaft. For some types of transmission, accurate control of the hydraulic system pressure is essential. In this case, a diaphragm-operated linear variable differential transformer (LVDT) has been used with some success experimentally. In production systems, a pressure sensor based on etched-silicon technology is likely to be the best device from the standpoint of cost and reliability.

2.3. SUSPENSION CONTROL

In a conventional automotive suspension, two incompatible requirements have to be reconciled by achieving a compromise between good vehicle handling characteristics and the provision of a comfortable ride. The conventional approach to suspension design uses steel springs to carry the vehicle body and dampers (shock absorbers) connected in parallel with the springs to absorb energy put into the system when the wheels encounter bumps in the road. With this approach it is possible to achieve good ride characteristics or good handling characteristics, but not both simultaneously.

In racing cars and other high-performance vehicles, a different approach has been used. The system adopted is known as an active suspension. With this approach either the vehicle springs or the dampers or both are replaced by controllable devices with variable characteristics. This allows the suspension characteristics to be altered to suit current driving conditions.

There are two major types of controllable suspension. In one approach, known as active suspension (developed particularly by Lotus), the vehicle springs are replaced or substantially augmented by hydraulic jacks. These jacks are electronically controlled to maintain the body of the vehicle as level as possible irrespective of the road surface being traversed by the vehicle. This approach can lead to an expensive system that requires large amounts of power to operate, and its use is normally restricted to racing vehicles.

An alternative (and much cheaper) approach is known as adaptive damping. In this system the stiffness of the shock absorbers is controlled, either continuously or in a series of steps, by an electronic control system that can react to bumps in the road, turning movements of the steering, and speed changes. The use of adaptive damping provides a soft, comfortable ride for low-speed straight-line driving, with a progressively harder ride and improved handling characteristics during high-speed driving and rapid turning movements. Because of the lower costs, the adaptive damping approach is the most likely to be used in production vehicles, and it is becoming common on more expensive vehicles.

Figure 1.9 shows the layout of a typical adaptive damping system. The most notable feature is the large number of sensors required. The gearbox speed sensor is usually already provided to pass information about the vehicle's speed to the transmission and engine control system. The throttle position sensor is also already provided for use by the engine control system. Suspension control requires further sensors for steering-wheel position and velocity, to inform the suspension control system when the driver initiates a turning maneuver. The most likely sensor for this purpose is an optical digitizer disk, with the digital information obtained being processed electronically to give information on both position and rotational velocity. Some prototypes used potentiometer sensors for steering-wheel movement sensing, but problems of wear were experienced, and the optical approach is more likely to succeed in practice.

At least one accelerometer is required to provide information on vertical acceleration of the vehicle body. Improved performance is obtained by using four accelerometers, one at each corner of the vehicle. In some systems these may be augmented by a further lateral accelerometer placed at the center of the vehicle. These accelerometers are likely to be constructed using piezoelectric or etched-silicon devices, perhaps combined with thick-film technology [15].

The measurement of static and dynamic wheel-to-body displacement poses particular problems because of the large ranges of both displacement and velocity to be measured. One promising method

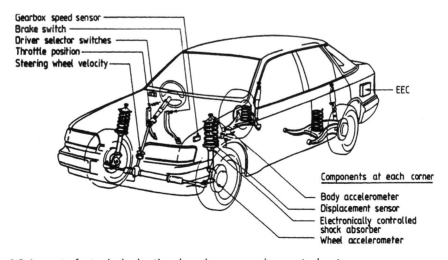

Figure 1.9. Layout of a typical adaptive damping suspension control system

is to use an inductive approach, in which the shock absorber piston moves vertically within a coil of wire. Effectively, this turns the shock absorbers into a simple form of LVDT [16]. Other workers have used linear potentiometers, normally fitted within the shock absorbers, but experience showed that this approach suffered from problems of noise and wear.

Finally, to achieve optimum performance, wheel accelerometers are also required, in addition to body accelerometers, to measure the vertical acceleration at the wheel hubs (i.e., on the unsprung side of the suspension). Once again, it is likely that these accelerometers will be constructed from micromachined silicon.

From this brief description, it can be seen that a diverse range of sensors is required to provide sufficient input information to actively and effectively control the vehicle suspension in response to all the various inputs from the road to the vehicle. In the longer term, embedded simulation may be used to allow the sensing requirements to be reduced. The target of this approach is suspension control using only one major vertical and one lateral acceleration measurement, together with steering-wheel motion input to give predictive information.

2.4. SENSORS FOR ANTILOCK BRAKING AND TRACTION CONTROL

The final major powertrain control system fitted to most vehicles is that concerned with antilock braking and antispin traction control. Antilock brakes work by sensing whether, on braking, one wheel starts to slow down towards a locked condition faster than the other wheel. If this is detected, hydraulic braking pressure is taken off that wheel, which is then allowed to speed up until it is again rotating at the speed of the other wheels. Braking pressure is then reapplied. Antispin or traction control is the exact reverse of antilock. It is intended to retain traction when a wheel spins on a slippery surface (such as ice). When a wheel starts to speed up towards spin, braking is applied to that particular wheel, and backing off the (electronic) engine throttle through the engine control system reduces engine power.

It should be clear from the above description that if a vehicle already has electronic engine control, with electronic throttle and electronic antilock braking, then the provision of an antispin system can be provided at almost no extra cost. This is a classic example of the benefits of integrating different electronic control systems around the vehicle.

The critical sensors involved in antilock/antispin systems are those used for sensing wheel rotation speed. A range of electromagnetic devices is currently used in production vehicles, including variable reluctance and Hall effect devices—both of which detect the passage of ferromagnetic teeth on a steel digitizer ring mounted on each wheel. Because of the exposed position and the inevitable contamination from salt water and mud, inductive or magnetic sensing techniques are essential. These sensors are unlikely therefore to change dramatically in the future. They are likely, however, to be among the first automotive sensors to benefit from the inclusion of "intelligence" at the sensor. At present the train of relatively high-frequency pulses from the wheel sensor is passed through the vehicle wiring system to a central control unit. This process is always at risk of electromagnetic interference in the severe automotive environment. The information actually required, however, is not wheel speed, but the *change* in wheel speed, and this is essentially a very low-frequency event in electronic terms. Sensing the change in rotation rate could easily be handled by a suitable smart sensor, and the data provided would potentially be much more suitable for transmitting over the multiplexed data buses that are already beginning to appear on many vehicles.

2.5. TIRE AND WHEEL SENSING

Vehicle tires and wheels are probably the last remaining major parts of the mass-produced motor vehicle that have not to date benefited from electronics. However, the wheels and particularly the tires of a road vehicle represent the final and probably the most critical link in the chain by which engine power is used to create vehicle motion. The tire is a key component in conventional ground vehicles. With the exception of wind resistance and gravity, all external forces are transmitted through the tire-to-road contact patch.

If sensors could be applied to the tires of a vehicle, a number of benefits might be obtained. Continuous monitoring of the coefficient of friction between the tire and road could enhance both vehicle safety and performance. Tire temperature could also be measured, and this has been shown to be a good predictor for tire blowouts. Work undertaken at Southampton University demonstrated that most high-speed tire blowouts result from prolonged running on under-pressure tires. The resultant heating can raise the tire temperature by up to 50 °C, and failure ultimately results from heat-induced softening of the tire material and structure.

The Southampton study also demonstrated that it is feasible to measure tire pressure while a vehicle is in motion. Once again, tire pressure can be used as an important indication of imminent failure. As noted above, most tire bursts are the result of prolonged running on under-pressure tires. It is a sad fact that most car drivers do not check their tires on a regular basis, and a dashboard tire monitoring system offers real safety advantages. In the case of both tire pressure and tire temperature, suitable sensors are readily available. The major problem in manufacturing tire pressure, temperature, or strain measurement systems is that of powering the measurement system, and recovering sensor signals from the rotating wheel at an economic cost. A number of different approaches have been used with varying degrees of success. Some manufacturers have incorporated batteries within the wheel hub to energize the measurement circuits, and have used radio telemetry to recover the data. Other studies have reported the successful use of inductive coupling across air gaps of up to 10 mm. The Southampton study made use of inductive coupling for both energizing the measurement circuit and for data retrieval. In that study (see reference 17) power was fed from the chassis to the wheel at a frequency of 1 MHz, and encoded data passed in the other direction through the same inductive link at a carrier frequency of 100 kHz.

A second study undertaken by the Southampton group made use of a purely mechanical sensor, in which pressure changes were used to alter the separation of a pair of permanent magnets. The magnet positions were sensed from the vehicle chassis by a Hall probe [18]. The device incorporated a mechanical temperature compensation system, and thus avoided any need to pass power to or retrieve data from the rotating wheel.

2.6. VEHICLE-BASED EXTERNAL SENSORS FOR DRIVER SUPPORT

The term used by the automotive industry for driver support equipment is Integrated Driver Support (IDS). The first commercial application of IDS is the autonomous cruise control (ACC) system, initially introduced on up-market vehicles. ACC uses a vehicle-based rangefinding system as described in the next section.

The exclusivity of IDS systems such as ACC is likely to be short-lived. Most European car manufacturers are developing the feature for vehicles based on high-volume platforms that will start to

Table 1.2. Summary of vehicle external environment transducer characteristics

System	Range	Resolution	Cost	Carrier frequency	Environmental Considerations	Distance/velocity	Beam width (approx.)
Pulse Doppler radar	Very long, over 500 m	Around 0.5 m	High	77 GHz	Very robust in harsh weather conditions	Velocity only	10°
Microwave radar	3–150 m	Around 0.5 m	Medium	60 GHz	Very robust in harsh weather conditions	Distance and velocity	10°
FMCW radar	2–100 m	2.8 m	Medium	77 GHz	Very robust in harsh weather conditions	Distance and velocity	10°
Active infrared (laser/LED -based)	LED 30 m (traditional) laser > 100 m	High	Medium/high	N/A	Will not work in strong sunlight	Distance	2°
Lidar	< 60 m for pedestrians	High	Medium/high	N/A	Some sensitivity to rain, fog, snow; very sensitive to dirt	Distance	2°
Passive infrared (PIR) imaging	Up to 25 m	400 µrad	Medium	N/A	Some sensitivity to rain, fog, snow; very sensitive to dirt	Distance	10–30°
Ultrasound	8 mm–20 m	5 mm	Low	20–50 kHz	Fur/furry-textured clothing absorbs signals; low repetition rate	Distance (and velocity if Doppler used)	25°
Image-based (cameras)	Up to 50 m	500 µrad	Medium/high	N/A	Strong shadows, poor lighting can cause problems	Distance	N/A
Capacitive	Up to 2 m	10 mm	Low	N/A	Sensitive to rain and snow and radio-frequency interference (RFI)	Distance	N/A

appear within the next two to three years. A number of Japanese and Korean manufacturers are also planning to include ACC on new models for both the domestic and export markets.

Other IDS systems likely to appear in the next decade are targeted at making the roads safer for both vehicle occupants and other road users. These include collision warning and collision avoidance systems, as well as sensors that trigger external protection devices, such as external airbags or side protection curtains fitted to Heavy Goods Vehicles (HGVs) to reduce or prevent injury to vulnerable road users.

2.6.1. Sensors for Adaptive Cruise Control (ACC) and Collision Avoidance

ACC makes automatic adjustments to the speed of a car to insure a constant headway is maintained between vehicles. As a consequence, traffic flow is smoothed and the more comfortable progress of the car reduces fatigue and stress for the driver.

Like a standard cruise control, ACC is capable of maintaining a cruising speed at a preset level, but is also able to adjust the speed to maintain a safe distance from the vehicle in front. This is possible through automatic control of the accelerator and brakes, which then allows pedal-free progress and so alleviates some of the work associated with long distance travel over motorways and urban expressways.

Quite apart from being seen as a convenience feature, ACC appears to be able to help reduce traffic congestion by keeping the traffic moving. Simulation flow tests in Sweden showed that if just one car in ten has ACC installed, traffic flow is dramatically improved.

ACC is the first of a generation of driver support systems that will ultimately lead to convoy driving, collision avoidance and perhaps even the "electronic chauffeur." Driver acceptance of the technology will evolve as the systems are brought to market, and there is little doubt that the benefits of IDS features in general will win wide recognition over the next decade or so.

The sensor requirements of ACC consist in the main of distance measurement (or rangefinding). Most of the systems that have been or are about to be launched use a radar sensor fitted to the front of the vehicle, often behind the vehicle registration plate, to measure the speed and closing distance of the vehicle in front. Frequency-modulated continuous wave (FMCW) radar is normally used, operating at a frequency of 77 GHz [19]. The first-generation ACC systems were found to suffer from poor angular coverage (typically 10°) and a near-range cutoff (typically 5 m), which limited their use in dense or slow-moving traffic. On even moderately curved roads, the narrow beamwidth necessitates some form of beam steering. However, these first-generation systems function well on motorways and urban expressways.

A number of laser-based optical rangefinding (LIDAR) systems for ACC have also been proposed [20,21]. Most of the proposed systems adopt a pulsed mode of operation and use infrared to achieve better penetration in mist or rain. The system described by Lissel and others is typical, and it has a wavelength of 1.55 μm and a pulse duration of 50 ns. This type of sensor has better resolution than the radar systems and is capable of detecting objects as small as 0.1 m at ranges of 100 m. However, unlike radar the performance of LIDAR systems, it is markedly degraded by dirt, mist, or rain, as well as road spray.

On a demonstration vehicle produced by Lucas-Varity and Thomson CSF, an additional refinement allowed the vehicle to accelerate as soon as the indicator or turn signal switch was activated. The project participants believed that this feature encouraged proper use of the indicators, in addition to assisting the vehicle in lane-changing maneuvers.

2.6.2. Sensors for the Vehicle Environment

The use of transducers to sense obstacles close to the vehicle is not restricted to ACC. A number of studies have proposed systems that can warn the driver of the presence of pedestrians or other obstacles during reversing, as well as monitor the "blind spot" during overtaking maneuvers. Visibility of the road around a vehicle has always been problematic even on bright and sunny days. A number of vehicle manufacturers are studying the idea of replacing mirrors with video cameras to eliminate blind spots. This approach has the added advantage of improving the car's aerodynamics, reducing wind noise, and increasing fuel economy.

2.6.2.1. ULTRASOUND

Ultrasound offers an inexpensive method by which the presence, distance, and direction of objects close to the vehicle can be measured. A number of manufacturers already offer ultrasonic parking/reversing aids, often using an audio signal for driver feedback. Ultrasonic distance measurement is normally based on a pulsed, time-of-flight approach, and distance can be measured to an accuracy of a few millimeters or better [22]. The resolution depends primarily on the precision with which the speed of sound is known, which varies with local air properties. The beamwidth of an ultrasound transducer is typically 20–30°, but the relatively low cost means that it is usually feasible to employ more than one device in a given application. The repetition rate is normally no more than 10 Hz because of the relatively low speed of sound compared to optical/radar pulses. Thus, ultrasound systems are restricted in their application to low-speed maneuvering and other applications where the relative velocity is low.

While most surfaces and objects are good reflectors of ultrasound, one study noted that pedestrians wearing fur or heavy wool coats provided little or no attenuation [23].

2.6.2.2. VISION AND IMAGE-BASED SYSTEMS

The image from a charge coupled device (CCD) camera consists of a two-dimensional picture generated in a form that may be stored and processed by computer. The images are typically produced at a 25 Hz frame rate [24] and are stored as a matrix in computer memory. The digitized image can then be processed, displayed, and searched for recognizable features using a model-specific filter. Maximum response locations are used as the initial search points for the model matching process, possibly supported by other features such as shape, symmetry, or the use of a bounding box. In principle, image processing techniques can be used to give information about the presence of humans at close ranges, that is, less than 0.3 m and up to 50 m. As humans vary a great deal in size and shape, image analysis systems find it difficult to distinguish between a tall adult standing 10 m away and a small child standing 2 m away. This could be overcome by combining image processing with a ranging system, for example, ultrasound.

It has been shown that by combining ultrasound and image processing, a typical CCD camera can give an accuracy of 1.5 mm for distance estimation, 2° for bearing, and a recognition rate or success rate of 90% in the presence of clutter. A combined system of this type was described in a study by Zhang and Sexton [25].

An alternative to the use of visible-light images is passive infrared (PIR). Infrared (IR) sensors use the fact that all objects with a temperature above absolute zero emit black body radiation. Different objects emit light of different wavelengths. The human body emits IR radiation in the wavelength region 8 µm–14 µm.

A PIR camera uses an array of detectors to image the radiation of a warm body, which can be significantly higher than the background temperature. In most commercial PIR devices, the radiation passes through antireflection coatings (to minimize the reflection losses caused by the big change of the refractive index from air to silicon) and is then absorbed by a detector fabricated from polyvinylidene difluoride-trifluorethylene, or P(VDF-TrFE).

Although PIR systems for automotive use are relatively expensive at present, they offer considerable scope for monitoring the area immediately in front of a vehicle where radar, ultrasound, and so forth are not effective. PIR sensors typically have a range up to 25 m and a field of view of 10–20°. The field of view can be increased, but this adversely affects the resolution.

2.6.2.3. CAPACITIVE PROXIMITY SENSORS

Capacitive sensors have been used for many years in industrial environments to detect the presence of nonmetallic objects and for liquid level measurement [26]. They have also been considered for detecting obstacles in low-speed maneuvers such as parking. Transducers of this type operate by sensing the change in dielectric constant that occurs when an object with dielectric properties different from those of air approaches a sensing electrode. Capacitive sensors can accurately measure distance up to a range of 2 m and can determine the size and shape of the detected object [27]. Capacitive sensors do not contain any mechanically moving parts, are insensitive to dirt, and will work in the absence of light. The associated circuitry is simple, and the devices are low cost, all of which appears to make them ideal candidates for use with automotive telematic systems. The fact that they have not (to the author's knowledge) been introduced despite extensive trials is most probably due to the fact that (like other high-impedance devices) capacitive sensors are prone to RFI.

2.6.2.4. SUMMARY TABLE

To conclude this section, table 1.2 is presented, which summarizes the characteristics of various forms of vehicle-mounted sensors for measuring the vehicle environment.

2.7. Vehicle-based Safety Sensors

Until about two decades ago, most in-vehicle safety systems were mechanically based. Two approaches to occupant protection were adopted: the use of harness restraint systems to prevent passengers from moving during a crash and a structural design approach in which energy-absorbing regions ("crumple zones") were incorporated within the vehicle body.

The introduction of airbags has brought a third element into play: the use of electrically fired devices to cushion the driver and other occupants. Airbags, together with fuel cutoff switches, battery disconnect systems, and seatbelt pretensioners, require sensors for their correct operation. These transducers sense the rapid deceleration associated with a crash, but must also discriminate between a crash and "legitimate" maneuvers, such as driving over curbs, hitting potholes, or heavy braking. Since the firing of an airbag is very much a "last resort" as far as safety systems are concerned, and also because airbag operation has occasionally been found to be hazardous, airbag sensors are relatively insensitive and require decelerations in excess of 20 g sustained for at least 20 ms before they are deployed. Figure 1.10 shows a typical set of airbag deceleration/duration requirements, from which it can be seen that the transducer has to respond to both high intensity, short-duration impacts as well as longer-intensity, lower-duration events. Essentially what is being transduced is the energy dissipated in an impact, which can be calculated from the integral of the acceleration-time function. Most crash sensors are therefore inertia switches. Since the required function is one of switching rather than deceleration measurement, conventional accelerometers are not normally used.

One common form of inertia switch consists of a weight, often in the form of a steel ball or cylinder, which is fixed to one end of a spring. The weight and spring are contained within a tube. When the assembly is accelerated or decelerated along its axis, inertia forces cause the weight to compress or extend the spring. The deflection of the weight is proportional to acceleration and is sensed by placing an appropriate sensor in the side of the tube. An alternative design, which has the advantage of being equally sensitive to any deceleration in a (normally horizontal) plane, uses a ferromagnetic sphere, retained at the bottom of a conical or dish-shaped "saucer" by a permanent magnet. A simplified form is shown in figure 1.11. Rapid deceleration causes the ball to break free from the magnet, and it rolls towards the rim of the saucer where it closes a switch.

Fuel cutoff and battery disconnect sensors operate in a similar fashion to airbag sensors, but are normally made more sensitive with switching thresholds in the range from 8–12 g. The likelihood of unintended operation during the vehicle's lifetime means that a somewhat more complicated mechanical arrangement is normally used in which motion of the seismic mass disturbs an over-center mechanism. Resetting the transducer requires external intervention—often through the medium of a reset button located in the trunk or elsewhere in the vehicle, which the driver has to press. An argument frequently advanced for this arrangement is that the requirement to open the trunk prior to resetting the fuel pump cutoff forces a driver to walk around the car following a minor collision, thus enabling him or her to

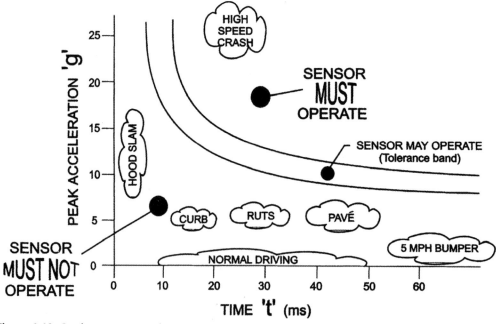

Figure 1.10. Crash sensor operation curve

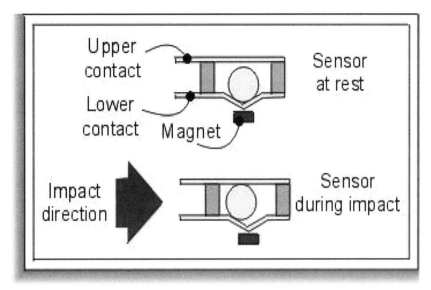

Figure 1.11. Crash switch (simplified)

observe any fuel leakage from damaged pipes or tank. It is noteworthy in this context that the frequency of postcrash fires has decreased markedly since the widespread introduction of fuel pump cutoff switches.

Other reported applications for crash switches include seat belt pretensioners, activation of hazard warning lights, mobile phone distress beacons, door lock releases, and antitheft devices.

Some crashes do not result in a deceleration severe enough to trigger inertia switches of the types described above, but they are nevertheless life threatening. Most incidents of this kind involve the vehicle turning over, and a range of rollover switches has been developed with the purpose of firing airbags, switching off fuel pumps, and so forth in the event of this type of accident occurring. Figure 1.12 shows a typical design, in which a ball is contained within a V-shaped enclosure that requires rotation to more than 60° before the device operates by breaking a light beam.

3. VEHICLE-HIGHWAY SYSTEMS

The major telematic products currently under development by the road transport industry include the following:

- Vehicles under some form of automatic or "telematic" control have to monitor others in their immediate environment, normally using some combination of radar, image processing, laser rangefinders, ultrasound, and vehicle-vehicle communications. The sensor and signal processing requirements for this task are very challenging in view of the low budgets available for on-vehicle equipment and the safety-critical nature of the task.
- Highway operators use a variety of sensing techniques to monitor traffic flow and detect incidents (breakdowns and accidents). In the near future, many countries are likely to implement

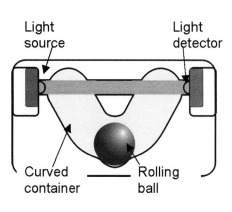 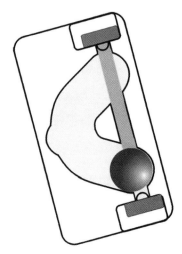

- V shape gives 60° trigger angle

- Light beam is broken by ball

- Output can be electronically filtered to improve false triggering

Figure 1.12. Rollover switch operation

  systems that charge for road use. Both applications may require on-board or infrastructure-based sensors for vehicle type, speed, weight, or even emissions.
- Many navigation and traffic congestion warning systems (e.g., the Trafficmaster system available in the United Kingdom) use highway sensors [28] to monitor and collect information, which is then used to optimize the driver's route choice in the light of prevailing traffic conditions. A variety of sensing techniques are used, including optical systems (e.g., IR speed sensors), Doppler microwave devices, image-based systems (using analysis of video pictures), as well as more conventional roadway-based transducers such as inductive loops.
- So-called drive-by-wire systems are becoming widespread in automotive engineering. This is not as big a step as may at first be supposed: Many of the direct mechanical links between the driver and the vehicle controls (such as the brakes, throttle, steering, etc.) have been supplemented by electronic connections for some time. Examples include diesel engine management systems, antilock braking (ABS), traction control systems, and cruise control—all of which can and do intervene in the vehicle's operation. A number of high-performance vehicles (such as the Jaguar XK8 [29]) make use of a drive-by-wire throttle. The reasons for making this change are cost and convenience: It makes the job of the vehicle designer much easier if large, fixed-geometry components (such as steering columns or brake and throttle cables) are replaced by wiring

harnesses, particularly in the crowded dashboard area. However, reliance on telematic drive-by-wire systems does add to the safety-critical nature of many vehicle functions.
- Convoy driving [30] is increasingly being seen as offering the most promising means of improving road usage (i.e., the number of vehicles per kilometer of highway). Convoy driving involves setting up electronic "trains" of vehicles, which run close together (i.e., with reduced headway) under autonomous control. Vehicles taking part in convoys will have to have better condition monitoring systems than at present, since the consequences of mechanical or electrical failure (such as a tire bursting) on a participating vehicle operating at a reduced headway could be catastrophic.

The engineers responsible for producing telematic systems that directly affect the vehicle controls must be extremely careful to ensure that their designs are fail-safe and fault-tolerant. It is curious that society appears to be happy to accept the fact that more than fifty thousand deaths a year are caused by human drivers in Europe alone—but if an electronic system were to be responsible for even one such death per year, it would cause an immediate outcry. To function safely and to be generally acceptable, therefore, vehicles that are even partly controlled by telematic systems are likely to be very heavily dependent on sensors. As we have seen, these have to be both very cheap and very robust.

The subject of highway (road and roadside) sensors for vehicle telematics is very large, and space allows only a short overview to be presented in this chapter. Vehicle detection systems have been used for many years to monitor and control traffic. The first pneumatic-tube–based vehicle detectors were used as long ago as the 1930s for traffic signal control, and some pneumatic systems are still in use today. However, induction loops were first introduced in the early 1960s and rapidly became the dominant detection technology [31].

3.1. LOOP DETECTORS

Loop detectors consist of one or more loops of wire buried in the road surface as shown in figure 1.13. The loop forms the inductive part of an LC resonant circuit. At resonance the fundamental frequency of oscillation is as shown:

$$f = \sqrt{\left(\frac{1}{LC}\right) - \frac{R^2}{L^2}}$$

Variation in the inductance L, the series resistance R or the capacitance C will cause a change in frequency. Loop detector circuits are *designed* to be relatively insensitive to small changes in R or C, and automatically tune to resonance within a prescribed frequency band. Two detection systems are used. Probably the least error-prone approach is one in which a phase shift between the loop oscillator circuit and a fixed reference frequency is detected. However, an alternative approach is used by some systems, which detect the change in frequency that occurs when a variation in loop inductance takes place. In either case the circuit output is monitored by a microprocessor-based system scanning a number of channels.

As a vehicle is introduced into the inductive loop in the road, three effects occur:

- Magnetic and eddy currents are induced in the vehicle, which oppose the main field of the coil and therefore *reduce* the loop inductance.
- The magnetic flux density increases because of the iron content of the vehicle, which *increases* the loop inductance.
- The loop capacitance increases due to the proximity of the vehicle.

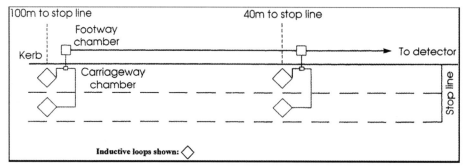

Figure 1.13. Two-lane loop detector layout

The eddy-current-induced inductance change is by far the greatest effect and is the one that makes this system of vehicle detection viable. The other two effects are usually neglected.

One problem with this system is that it can be unreliable for detecting bicycles. The low metallic content of a bicycle means that only small eddy currents are induced in the cycle's structure. If a cyclist crosses the loop wires at right angles to the direction of travel, eddy currents will only be induced across the width of the bicycle, and the resulting inductance change is normally too small to be detected. If bicycle detection is required, loops must be installed at an angle to the cyclist's direction of travel. This results in the production of two orthogonal magnetic vector components, one of which opposes the loop's magnetic field and is large enough to ensure bicycle detection. Bicycles with small wheels and/or nonmetallic frames can be very hard to detect; however, and the problem is not easily solved by increasing detector sensitivity since this often causes an increase in false detection signals.

3.2. INFRARED DETECTORS

Infrared (IR) detectors have been used as an aboveground vehicle sensor that avoids some of the problems (such as bicycle sensing) inherent in loop detectors. IR detectors may operate in active or passive modes, but whichever is used the essential difference between these devices and the inductive loop systems described earlier is that IR detectors can only detect *movement* within their zone of coverage, while loop detectors sense the *presence* of a vehicle.

Passive infrared (PIR) detectors consist of an optical system that focuses infrared energy onto a transducer. The transducer incorporates one or more ceramic pyroelectric elements behind a silicon window and is sensitive to radiation in the 6 to 14 μm region of the electromagnetic spectrum. This waveband corresponds to the emissions peak from targets such as motor vehicles and people. The detection algorithms used rely on sensing changes in radiation intensity as targets move in and out of the detection zone.

In contrast, active infrared detectors operate by transmitting encoded beams of infrared light onto a region of the road, normally from an overhead gantry. Reflections from vehicles traveling through the illuminated zone, which is limited to a small area, are focused onto a sensor matrix. Real-time processing of the output signal and comparison with the encoding data are used to indicate the presence of a vehicle in the target area.

Active IR detectors can also be used for speed measurement as shown in figure 1.14. The normal arrangement is for two detection zones per lane to be used. The time at which a target vehicle enters each zone is noted, and the vehicle speed can then readily be calculated as long as the installation geometry is known.

3.3. MICROWAVE DETECTORS

Microwave vehicle detectors (MVDs) use the microwave part of the electromagnetic spectrum. They operate by radiating a microwave beam in the X (10.587 GHz) or K (24.2 GHz) frequency bands. When the beam strikes a moving metallic object, a reflection is returned at a slightly different frequency (the Doppler effect). As the MVD relies on detecting this Doppler shift, it cannot detect stationary or very slow-moving vehicles. For an X-band system detecting a vehicle travelling at 50 kph, the Doppler shift is of the order of 1 kHz.

Some studies of MVDs have shown that their performance can be erratic. In work described in reference 31, it was shown that some MVDs failed to detect around 3.5% of all vehicles, and that heavy goods vehicles (HGVs) were detected an average of two seconds earlier than cars. This could lead to timing problems for traffic signal controllers.

3.4. PIEZOELECTRIC DETECTORS

Piezoelectric (PE) cables have been used to provide an accurate axle (rather than vehicle) detector for vehicle classification and speed surveillance applications. Piezoelectric cable uses the well-known piezoelectric effect [32], in which lateral compression of a cable made from a piezoelectric plastic material such as PVdF (Poly-vinylidene fluoride) results in the generation of a small charge that can be amplified and detected. In use, the piezoelectric cable is placed in a precut slot in the carriageway so that vehicles drive over it. After conditioning, the output usually takes the form of a pulse corresponding to each axle. PE systems of this type are very sensitive, allowing detection of vehicles such as motorcycles and

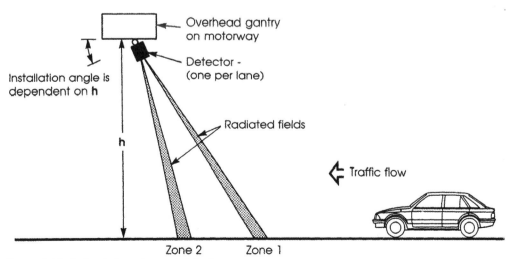

Figure 1.14. Active infrared speed detector

slow-moving bicycles. By measuring the time interval between pulses from consecutive piezo-detectors, it is possible to measure speed accurately in addition to axle counting.

PE detection systems have the disadvantage that they cannot provide any information about lane occupancy, since outputs are only provided as a wheel crosses the PE cable.

3.5. VIDEO IMAGE PROCESSING

Although not strictly a sensing system, video image analysis is included in this section for completeness. Video image processing techniques are increasingly being applied to detect the presence and speed of vehicles in real time, as well as for vehicle identification through number-plate recognition. A great deal of research is currently underway in this area, principally because the rapidly falling cost of image-analysis hardware is now making it feasible to design systems for highway use. Existing systems range from relatively low-level devices in which an image is divided into a number of windows, and changes to the "scene" in each window are assumed to indicate the presence of a vehicle, to more sophisticated approaches in which vehicle types and number plates are automatically identified and classified.

4. CONCLUSIONS AND FUTURE DEVELOPMENTS

In conclusion, this chapter has attempted to show the breadth of sensing and transduction techniques currently in use for automotive telematics. From the preceding discussion it can be seen that the field of automotive engineering is very active, with research pushing forward on several fronts. Drive-by-wire systems are in the late stages of development by a number of vehicle builders and component manufacturers, and it is almost certain that during the coming decade many or all of the mechanical links between driver and vehicle will be partially or completely replaced by electrical connections. The electronic systems currently incorporated in a motor vehicle account for 15%–20% of the cost of building the vehicle, and industry sources estimate that by 2010 this will rise to more than 30%. A large part of this increase will be the result of increased sensor use (and the associated signal conditioning systems).

Although few new sensing techniques are likely to be developed for automotive or highway use, sensor fabrication methods continue to develop as part of the drive to reduce automotive manufacturing costs. The advent of silicon micromachining techniques has meant that much greater intelligence can be incorporated within a sensor, with many of the signal conditioning and compensation circuits being fabricated on the same wafer as the sensor. Micromachining has also brought about improvements in reliability, principally as a result of the reduction in the number of device interconnections. Another fabrication technique finding automotive applications is the use of thick-film processing [33]. Thick-film techniques, while leading to larger package sizes than those produced by silicon methods, are cost effective at much smaller volumes (thousands or even hundreds of devices against hundreds of thousands for silicon).

It is interesting to note that, in many cases, the sensor technology required for vehicle telematics already exists, or at least can easily be developed from today's systems. Most of the problems that have to be addressed before vehicle telematics can be widely introduced center on legal issues (such as liability in the event of an accident), customer acceptability, or unwillingness on the part of vehicle manufacturers or highway authorities to make the necessary investment. Part of the problem has to do with the perception of risk. Considerable effort will have to be put into persuading drivers that surrendering some of their authority over a vehicle is acceptable. For some reason society tolerates the fact that human drivers cause several hundred deaths a month in Europe—but there would be an enormous outcry if it were found that telematic systems were causing even one fatality a year!

In present-day vehicles the most likely consequence of a serious electronic systems failure is that the engine stops. While there is no doubt that this can be dangerous, in most circumstances it is merely inconvenient. The drive to increase road usage, however, combined with the rise in traffic levels, implies that driving will become increasingly automated (or "telematic"). In the process more and more sensors and monitoring systems will be needed as the electronics on a vehicle become increasingly safety critical. The engineers responsible for producing telematic systems that directly intervene with the vehicle controls must, therefore, be extremely careful to ensure that their designs are fail safe and fault tolerant. Systematic techniques are needed to produce technology that "degrades gracefully," to use an industry expression. In software this aim is partly achieved by the use of Lean Formal Methods (LFM). However, similar formal design techniques will also be required for safety-critical hardware systems, and hardware/software interactions will require particularly close scrutiny.

Convoy driving (perhaps using specially designated highway lanes) is under active consideration in many countries as a means of improving the traffic density and usage of motorways. Trials of the technology needed to achieve convoy driving have been undertaken in the United States, Europe, and Japan. The aim is to improve the traffic flow and density through the use of vehicle-to-vehicle communications. Convoys of electronically linked vehicles travelling with reduced headway are expected to significantly improve road usage and reduce fuel consumption, journey time, and driver stress.

Convoy driving systems are the logical extension of the Autonomous Cruise Control (ACC) systems now being deployed. However, it seems likely that safety concerns will require vehicle-based sensing to be backed up by vehicle-to-vehicle data communications links. For convoy driving to be fail safe, each vehicle will have to communicate its status (such as speed, intended lane changes, etc.) to its neighbors.

In theory, elimination of at least part of the human response from the control loop should allow vehicles to travel closer together than at present (headway reduction) with improved safety. The introduction of telematics systems such as ACC brings the advent of convoy driving much closer. However, the safety of convoy driving in the event of full or partial systems or mechanical failure on one of the convoy vehicles has not been examined in depth, and further research in this area is urgently required.

Highway systems are also becoming more safety critical. Real-time traffic control is already being undertaken, and, in some parts of the United Kingdom, variable speed limits have been in place for several years [34]. Consideration is currently being given to passing these speed limits to vehicles, initially for driver assistance in the form of a dashboard display. However, the possibility of direct control of vehicle speeds by the highway authorities is also being investigated, and the safety implications of failure in such a system are clear. It is likely that enhanced sensing and monitoring systems will be required if current patterns of vehicle use are to be maintained. It seems clear that the use of sensors and measurement systems on both road vehicles and highways will increase considerably during the next decade.

REFERENCES

1. Hojer M. (1998, November). Transport telematics in urban systems: A backcasting Delphi study. *Transportation Research Part D: Transport and Environment* 3(6): 445–63.
2. Westbrook, M. H., and J. D. Turner. 1994. *Automotive sensors.* Bristol: IOPP.
3. Ibid., chap. 3.

4. Baney, W., D. Chilcott, X. Huang, S. Long, J. Siekkinen, D. Sparks, and S. Staller. 1997. Comparison between micromachined piezoresistive and capacitive pressure sensors. *Society of Automotive Engineers (SAE) Special Publications* 1311 (Nov.): 61–64.
5. Westbrook, M. H., and J. D. Turner. 1994. *Automotive sensors*. Bristol: IOPP. Chapter 2.
6. Hillier, V., and P. Coombes. 2004. *Fundamentals of motor vehicle technology*. Cheltenham, UK: Nelson Thornes.
7. Cartnell, B. C., and F. L. Zeisler. 1988. An engine mass air flow meter. SAE paper 760017.
8. Joy, R. D. 1989. Air flow for engine control. SAE paper 760018.
9. Cockshott, C. P., and J. P. Vernon. 1983. An air mass flowmeter for test cell instrumentation. In *Proceedings of International Automotive Electronics Conference*, 20–26. IEE publication 229. London: IEE.
10. Hetricj, R., W. Fate, and W. Vassell. 1982. An oscillatory mode oxygen sensor. *IEEE Transactions on Electron Devices* 29: 129–32.
11. Haaland, D. 1990. Internal-reference solid electrolyte oxygen sensor. *Analytical Chemistry* 49: 1813–17.
12. Velasco, G., J. Schnell, and M. Croset. 1984. A thin solid-state electrochemical gas sensor. *Sensors and Actuators* 2: 371–84.
13. Anastasia, C. M., and G. W. Pestana. 1995. A cylinder pressure sensor for closed-loop engine control. SAE paper 870288.
14. Richardson, R. M., J. J. Main, and J. Lindre. 1988. A five-speed microprocessor controlled economy transmission. In *Proceedings of the 4th International Conference on Automotive Electronics*, 32–38. London: IEE.
15. Sion, R. P., and J. D. Turner. 1992. The design of a novel accelerometer for automotive applications using thick-film transduction. In *Proceedings of the 9th International Conference on Automotive Electronics*. London: I. Mech. E.
16. Atkinson, J. K. 1989, May. An intelligent suspension damper employing distributed processing. In *Proceedings of the 20th ISATA Conference*, Firenze, Italy.
17. Hill, M., P. Malson, and J. Turner. 1990. The development of a low-cost system for monitoring tyre pressure. *IEE Colloquium on Chassis Electronics Digest,* Mar. 23, no. 1990/047.
18. Hill, M., J. Turner, and H. Sadek. 1992. The development and testing of a car tyre pressure sensor. *Proceedings of 25th ISATA Conference*.
19. Russel, M., A. Crain, R. Campbell, A. Clifford, and W. Miccioli. 1997. Millimetre wave radar sensor for automotive intelligent cruise control. *Transactions of the IEEE on Microwave Theory and Techniques* 45(12): 2444–52.
20. Krishnaswami, K., and M. Tilleman. 1998. Off the line-of-sight laser radar. *Applied Optics* 37(3): 564–65.
21. Lissel, A., R. Bergholz, and R. Holze. 1995. Automatic Distance Regulation (ADR): A new system for driver assistance in future VW vehicles. In *Proceedings of the Institution of Mechanical Engineers Autotech Conference*, paper C498/8/186/95.
22. Webb, P., and C. Wykes. 1996. High resolution beam forming for ultrasonic arrays. *IEEE Transactions on Robotics and Automation* 12(1): 138–46.
23. Chou, T. N., and C. Wykes. 1997. An integrated vision/ultrasonic sensor for 3D target recognition and measurement. In *IEE Sixth International Conference on Image Processing and Its Applications*, 189–93. London: IEE.
24. Bertozzi, M., and A. Broggi. 1998. GOLD: A parallel real-time stereo vision system for generic obstacle and lane detection. *IEEE Transactions on Image Processing* 7(1):62–81.
25. Zhang, X., and G. Sexton. 1995. A new method for pedestrian counting. In *Proceedings of Image Processing and Its Applications Conference*. IEE Conference Publication no. 410, 208–12. London: IEE.
26. Seippel, R. G. 1993. *Transducers, sensors and detectors*. Reston, VA: Reston Publishing Company.
27. Karlsson, N. 1997. Capacitive detection of humans—a robot safety application. *Linkoping Studies in Science and Technology*. PhD thesis no. 616, Department of Physics and Measurement Technology, Linkoping University, S-581 83, Linkoping, Sweden.

28. Stevens, A., and D. K. Martell. 1993. Development and evaluation of the Trafficmaster driver information system. In *Proceedings of IEEE-IEE Vehicle Navigation and Information Systems Conference,* 251–58. Piscataway, NJ: IEEE.
29. Proceedings of the 1996 IEE Colloquium on the Electrical System of the Jaguar XK8. *IEE Colloquium Digest,* Oct. 18, no. 281: 12–21.
30. Hochstaedter, A., and M. Cremer. 1997. Investigating the potential of convoy driving for congestion dispersion. In *Proceedings of the IEEE Conference on Intelligent Transportation Systems,* 735–40. New York: IEEE.
31. *Vehicle detection systems and installation methods.* 1991. Cardiff: Highways Directorate, Welsh Office.
32. Turner, J. D., and M. Hill. 1999. *Instrumentation for engineers and scientists.* Oxford: Oxford University Press.
33. White, N. M., and J. D. Turner. 1997. Thick-film sensors: Past, present and future. *Measurement Science and Technology* 8: 1–20.
34. Harbord, B., and J. Jones. 1996. Variable speed limit enforcement—the M25 Controlled Motorway pilot scheme. *IEE Colloquium Digest,* Nov. 18, no. 252: 5/1–5/4.

CHAPTER 2

Automotive Pressure Sensors

M. J. Tudor
S. P. Beeby

1. INTRODUCTION

Today, an average car utilizes more than fifty sensors, and a luxury car more than one hundred. Of these, one-third are currently based on micromachined sensors [1] with the proportion expected to increase in the future. Pressure sensors make up a significant proportion of this (around 15% by value) with manifold absolute pressure sensors being ubiquitous and tire and brake control pressure sensors becoming widespread.

Typical under-hood environments may involve temperatures from −40 °C to 1000 °C with thermal cycling, pressures up to 2000 bar, high humidity, salt spray, extreme shock loads, and high levels of vibration. Sensors must last for the lifetime of the car, which can be 15 years, at low cost per sensor in high volumes.

This chapter describes the main technologies applied to pressure sensing for use in the automotive environment. Section 2 discusses the primary types of pressure sensors categorized by fabrication technology. This covers, in turn, silicon micromachined sensors, piezoelectric sensors, screen printed thick-film sensors, thin-film sensors, optical sensors, and, finally, conventionally assembled sensors. Section 3 outlines the main approaches taken to extend the operating temperature of pressure sensors above 125 °C. This covers silicon on insulator and silicon on sapphire technologies as well as piezoelectric and optical approaches. Finally, in section 4 the major automotive application areas and primary technology used are discussed. This section covers inlet manifold pressure sensors, brake fluid pressure sensors, barometric pressure sensors, exhaust gas recirculation pressure sensors, engine oil/fuel pressure sensors, transmission pressure sensors, air conditioning pressure, coolant pressure, tire pressure, and cylinder pressure.

2. TYPES OF PRESSURE SENSORS SEGMENTED BY FABRICATION TECHNOLOGY

2.1. INTRODUCTION

Most pressure sensors work on the principle that the pressure is applied to a force summing device, which responds to the pressure, and this response is then measured to produce an electrical output. Originally, a widely used force summing device was the bourdon tube either in helical or spiral form. Alternatively, a bellows structure is often used. In more recent years diaphragms have become widespread, since they can be easily batch manufactured using micromachining techniques. These can be flat, bossed, or corrugated.

2.2. SILICON MICROMACHINED

2.2.1. Introduction

The measurement of pressure is a mature application of microelectromechanical systems (MEMS) technology and is one of its most successful areas. Research in this field began as far back as the 1960s, [2–4] and since then there have been many developments in micromachining processes and materials technology. MEMS pressure sensors utilizing a range of sensing techniques such as piezoresistive, capacitive, resonant, and optical have matured into a commercially successful solution for many sensing applications. The mechanical sensor element is typically (but not exclusively) a micromachined diaphragm. This section commences with an analysis of circular and rectangular silicon diaphragms and will introduce the basic micromachining processes and common techniques employed in the fabrication of silicon diaphragms. The mechanical properties of silicon diaphragms are briefly discussed, and the various sensing techniques used in automotive pressure sensors are introduced. Finally, real examples are discussed in section 4. The different sensing principles employed in automotive silicon pressure sensors to date will be described and illustrated.

The suitability of MEMS to mass-produced, miniature, high-performance sensors at low cost has made this technology extremely attractive to the automotive industry. Micromachined inertial sensors are commonly employed in automotive applications, including adaptive suspension, navigation, and safety applications such as crash detection for airbag deployment. Micromachined pressure sensors have been deployed in almost every automotive application, including inlet manifold, cylinder, brake fluid, tire pressure, and fuel injection. Automotive applications place challenging requirements on the silicon sensor, and both the design of the silicon component and its packaging require very careful consideration.

The diaphragm is used in both traditional and MEMS technology pressure sensors. The diaphragm is essentially a flexible flat or corrugated plate clamped around its edges. As pressure is applied the diaphragm will deform, the magnitude of the deflection being proportional to the pressure. The pressure is determined by measuring the amount of deflection or the strain induced in the surface of the diaphragm. The diaphragm structure can be fabricated using a range of bulk and surface silicon micromachining processes described below.

2.2.1.1. ANALYTICAL MODELING OF MICROMACHINED DIAPHRAGMS

Diaphragms are one of the few MEMS structures that can be modelled analytically. Bulk anisotropically etched diaphragms will be rectangular or square, and this is the most common type of micromachined diaphragm structure. The characterizing equations for a rectangular diaphragm with rigidly clamped edges and small deflections are given below.

$$y_o = \alpha \left(\frac{Pa^4}{Eh^3} \right)(1-\upsilon^2) \qquad (2.1)$$

$$\text{stress } \sigma = \beta \left(\frac{Pa^2}{h^2} \right) \qquad (2.2)$$

In equations 2.1 and 2.2, y_o is the maximum deflection at the diaphragm center under uniform pressure P, h is the diaphragm thickness, a is the length of the shorter side, and E and υ are the Young's modulus and Poisson's ratio, respectively. For a rectangular diaphragm, the coefficients α and β depend upon the ratio of the lengths of the diaphragm sides and the position of interest (e.g., edge or center of diaphragm). Assuming a square diaphragm, α equals 0.0151, and β equals 0.378 for the maximum stress that occurs along the edge of the diaphragm and 0.1386 for the maximum stress at the center of the diaphragm.

For circular diaphragms, which can be formed by isotropic bulk processes or surface micromachining, and assuming deflections of no more than 30% of the diaphragm thickness, the following equations apply. The deflection, y, at radial distance r of a round rigidly clamped diaphragm under a uniform pressure P, is given by equation 2.3,

$$y = \frac{3(1-\upsilon^2)P}{16Eh^3}(a^2 - r^2)^2, \qquad (2.3)$$

where a is the radius of the diaphragm. The maximum deflection, y_o, will occur at the diaphragm center where $r = 0$. Assuming $\upsilon = 0.22$ for silicon, the maximum deflection is given by equation 2.4:

$$y_o = \frac{0.178 Pa^4}{Eh^3} \qquad (2.4)$$

The deflection of a rigidly clamped diaphragm is shown in figure 2.1. It is also useful to provide an analysis of the stress distribution across a pressurized diaphragm in order to calculate electromechanical effects, which are used to measure the stress/strain induced in the diaphragm material (e.g., change in resistance of doped piezoresistors). The stress distribution varies across the radius and through the thickness of the diaphragm. For example, the neutral axis (shown in figure 2.1) experiences zero stress while the maximum stress levels occur at the outer surfaces. At a distance r from the center of the diaphragm, one face will experience tensile stress while the other experiences compressive stress. There are two stress components associated with a circular diaphragm: radial and tangential. The radial stresses are those in a direction from the center to the outer circumference (analogous to the spokes of a wheel). Radial stress, σ_r, at distance r from the center of a circular diaphragm is given by equation 2.5. The maximum radial stress occurs at the diaphragm edge ($r = a$) and is given by equation 2.6:

$$\sigma_r = \pm \frac{3}{8} \frac{Pa^2}{h^2} \left[(3+\upsilon) \frac{r^2}{a^2} - (1-\upsilon) \right] \qquad (2.5)$$

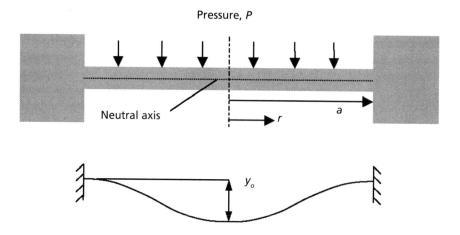

Figure 2.1. Deflection of a rigidly clamped diaphragm

$$\sigma_{r_{max}} = \pm \frac{3}{4} \frac{Pa^2}{h^2} \quad (2.6)$$

The tangential stresses are those in a direction parallel to the circumference. The tangential stress, σ_t, at distance r from the center of the diaphragm is given by equation 2.7. The maximum tangential stress occurs at the diaphragm center ($r = 0$) and is given by equation 2.8:

$$\sigma_t = \pm \frac{3}{8} \frac{Pa^2}{h^2} \left[(3\upsilon+1)\frac{r^2}{a^2} - (1+\upsilon) \right] \quad (2.7)$$

$$\sigma_{t_{max}} = \pm \frac{3}{8}(1+\upsilon)\frac{Pa^2}{h^2} \quad (2.8)$$

The dynamics of a micromachined diaphragm can be adequately characterized by linear plate theory. The undamped resonant frequency, f_n, of a clamped square diaphragm of uniform thickness and assuming a homogenous material of density ρ is given by equation 2.9 [5]:

$$f_n = 1.654 \left[E/\rho(1-\upsilon^2) \right]^{1/2} h/a^2 \quad (2.9)$$

The amount of damping present depends not only on the diaphragm design but more often on its packaging and surroundings. Where the micromachined diaphragm is separated from the pressurized media by a barrier diaphragm and hydraulic fluids are used to transmit the pressure to sensor, the frequency response of these components must also be considered.

Bossed and corrugated silicon diaphragms can also be fabricated using both anisotropic and isotropic etching. Such structures, however, cannot be modeled analytically and are typically simulated using finite element (FE) techniques [6]. Sandmaier has presented a set of analytical equations enabling basic optimization of diaphragm design [7], and corrugated silicon diaphragms have been discussed in the papers by van Mullem [8] and Jerman [9].

2.2.1.2. SILICON MICROMACHINING TECHNIQUES

The development of MEMS is based upon the ability to micromachine a variety of microscale mechanical structures from a silicon substrate. These micromachining processes have in many instances evolved from the development of microelectronic processes, and MEMS has benefited enormously from technologies developed by the semiconductor industry. For example, patterns and geometries are defined on the wafer surface using the same photolithographic techniques used in microelectronics fabrication. This involves a light-sensitive polymer (photoresist) being spun on the surface of the wafer and selectively illuminated through a patterned photomask. Silicon is an extremely good mechanical material. Having a Young's modulus approaching that of stainless steel, it is harder than iron and less dense than aluminum [10]. It also offers the attractive possibility that electronic circuits can be integrated on the sensor chip. There is a wide range of micromachining processes utilized in the fabrication of pressure sensors. These processes are typically divided into two categories: bulk and surface.

2.2.1.2.1. Bulk

Bulk micromachining refers to processes that remove material from the substrate thereby forming the mechanical structure from the bulk substrate material. A range of etching processes are available for removing silicon, the most common type being anisotropic wet silicon etching. Wet processes utilize liquid chemical etchants, and the term anisotropic refers to the fact that many of these etches attack the silicon at a different rate according to the crystalline direction. A typical example of such an etchant is potassium hydroxide (KOH), which etches the (100) planes 400 times faster than the (111) planes. When wafers are oriented in the (100) plane, this etch produces a rectangular diaphragm with sloping side walls that follow the (111) planes. A cross section of an anisotropically etched diaphragm is shown in figure 2.2. Since the crystallography of silicon wafers is well-known and precisely controlled, anisotropic etch processes provide repeatable diaphragm dimensions provided the etch depth is well controlled. This etch depth can be controlled by using a well-managed, timed etch process or, more preferably, by using an etch stop layer such as the buried oxide present in silicon on insulator (SOI) wafers or electrochemical etch stops.

Areas of the wafer where no etching is required are protected by a passivation, or masking layer. The masking layer material must be highly resistant to the particular etchant being used. In the case of

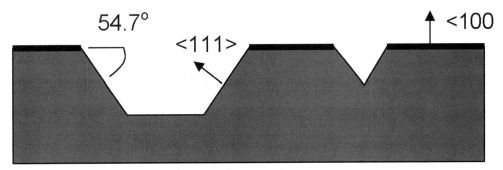

Figure 2.2. Anisotropically etched silicon diaphragm and groove

wet silicon etches, these layers are typically thermal silicon dioxide (SiO_2) or deposited silicon nitride (Si_3N_4). These passivation layers must first be etched to define the pattern of the diaphragms across the surface of the substrate. This is typically done using a dry etch process as discussed below. Wet etching is a batch process capable of producing hundreds of devices simultaneously across a group of wafers. In addition to anisotropic silicon etching, the isotropic wet etching of silicon dioxide (SiO_2) and aluminum is widespread.

Material can also be removed from the bulk of the wafer using "dry" etching processes that utilize plasmas to etch the silicon. Dry etch processes typically utilize a combination of chemical etching (chemically attacking molecular bonds) and physical etching (bombardment of the substrate) to remove material. The degree of anisotropy will depend upon the nature of the plasma used and the ratio of physical to chemical etching. The material selectivity (the difference in etch rates of different materials under the same etch conditions) of dry etching is greater than for wet etch processes, and therefore a much wider range of masking layers can be used. Patterned photoresist is able to withstand a wide range of dry processes and can therefore be used as the masking layer for a variety of etches. Plasma etching offers a much higher degree of control compared with wet etch processes and, although etch rates may vary across a wafer, is certainly the preferred etch process in microelectronic applications. Recent advances have led to a deep reactive ion etch (DRIE) process in which silicon can be etched with vertical sidewalls (90° ±1° typical) and very high aspect ratios (40:1 height to width of features). This is a very attractive micromachining process and is becoming increasingly employed in the fabrication of MEMS.

Bulk micromachining is widely used in the fabrication of pressure sensors, since the dimensions required in typical pressure sensing applications are compatible with the capability of the processes. The ability to fabricate the diaphragm from single-crystal silicon also offers important advantages given its excellent mechanical properties and inherent piezoresistivity (discussed below). The main drawbacks of bulk micromachining relate to the larger size of devices (and therefore fewer devices per wafer) and the reduced compatibility with microelectronics processes, which can make the integration of electronics on the sensor die more difficult.

2.2.1.2.2. Surface

Surface micromachining refers to processes whereby the mechanical structure is formed upon the surface of the substrate. This approach relies on a number of material layers being deposited and patterned on the wafer surface to make up the structure. The simplest form of surface micromachining is to use a sacrificial layer, as shown in figure 2.3. A layer of one type of material (e.g., SiO_2) is deposited onto the substrate and patterned. A different material is then deposited on top of the substrate covering the first layer, and this is also subsequently patterned. Finally, the first sacrificial material is etched leaving freestanding structures formed from the second layer material. The process requires good selectivity of the etch used to preferentially remove the first material and a method of depositing high-quality materials with low levels of built-in stress. Deposition process parameters must be tightly controlled and annealing steps may be required to minimize the stress. The sacrificial etch process typically involves the use of a wet etchant, and the capillary forces arising from the evaporation of the rinsing liquid (deionized water) can draw the mechanical structures down into contact with the substrate (a process known as stiction), after which structures can remain stuck to the substrate.

Typically, the sacrificial material is SiO_2 and the mechanical material polysilicon. This process can become quite complicated with several alternating layers of polysilicon and SiO_2. Material layers are deposited using a chemical vapor deposition (CVD) process. This process involves the reaction of

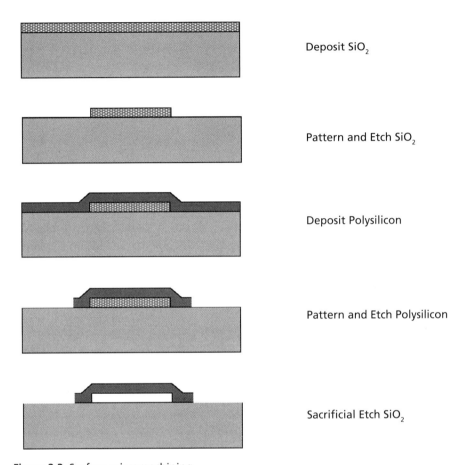

Figure 2.3. Surface micromachining

gaseous species at the substrate surface and may be carried out at low pressure (LPCVD), atmospheric pressure (APCVD), or with plasma enhancement (PECVD).

Diaphragms can be relatively easily fabricated using a surface micromachining process, but this is complicated by the need to etch the sacrificial layer. This problem can be overcome by sealing the diaphragm, using a subsequent deposition step or by bulk etching access holes through the substrate from the back of the wafer. Surface micromachined diaphragms are typically much smaller in size than their bulk machined counterparts, and the processes used are inherently compatible with integrated electronics. The main drawback is the inferior mechanical properties of the polysilicon layer, which often vary from batch to batch due to process fluctuations. The use of SOI wafers can enable structures to be fabricated from single-crystal silicon by using the buried oxide as the sacrificial layer.

2.2.1.2.3. Bonding

The bonding of substrates together is another key micromachining process widely used in both the fabrication and packaging of pressure sensors. Bonding processes enable first-order packaging to be carried out at wafer level, thereby reducing the cost and complexity of subsequent packaging steps. A variety of bonding processes exists, the suitability of each depending on the materials involved, the requirements of the application (e.g., hermetic seals), and temperature limitations. Hermetic seals are essential, for example, in the fabrication of absolute pressure sensors, since a reference vacuum must be maintained on one side of the diaphragm. In all the bonding processes, cleanliness is of fundamental importance, since particulates lead to voids, which weaken bonds and spoil seals. The stress created by the bonding process also needs to be considered, since this can affect the behavior of the sensor. This is especially true when dissimilar materials are bonded together and the bonding process is at elevated temperatures. The relaxation of bonds and any built-in stress over time is also an important consideration since this will lead to long-term drift in the sensor output.

Anodic bonding (also known as electrostatic bonding) is a commonly used process to bond silicon to glass. The glasses involved are typically Pyrex 7740 or Schott Borofloat 33, both of which have similar thermal expansion coefficients to silicon. The bonding method employs a combination of electrostatic fields and elevated temperatures. The process involves applying a negative potential to the glass wafer and heating the assembly up to typically > 350 °C. The magnitude of the applied potential will depend upon the thickness of the glass and the thickness of any oxide present on the silicon wafer (up to 100 nm being permissible). The resulting electrostatic field draws the wafers into intimate contact, which, combined with the high temperature, results in a perfectly hermetic chemical bond. Narrow metal tracks on the silicon can also be used for feedthroughs without compromising hermeticity. This process can also be used for silicon to silicon bonding provided a thin-film layer of glass is evaporated or sputtered onto one surface. Despite the similar thermal expansion coefficients, some residual stresses will result from this process.

Silicon fusion bonding is nowadays a routine process used in the fabrication of SOI wafers and MEMS alike. This technique does not require any intermediate material layers, and therefore the residual stresses trapped within the bonded pair are very low. The process requires very particular surface preparation. Both surfaces must be very clean, flat (polished), and hydrated. Two wafers are joined together at room temperature and will immediately stick due to van der Waals forces. The bond is completed by heating the assembly to temperatures typically in excess of 800 °C. These bonds are fully hermetic and low stress, but the high temperature required will not be compatible with some processes or materials.

Other bond types include using a silicon gold eutectic, which melts at 363 °C, forms a liquid layer, and then sets upon cooling. A similar bonding process can be achieved with a screen-printed glass frit, which typically melts at between 350 °C and 450 °C. Both of these processes are hermetic and strong but do result in high levels of residual stress. Nonhermetic weaker bonds can be achieved by using polymer adhesives. They can be spun onto wafers or automatically dispensed. The flexible nature of the adhesives can be of use in relieving other unwanted packaging stresses.

2.2.1.2.4. Packaging

The packaging of MEMS is a fundamental process and is often overlooked to the cost of the developer. Packaging costs can easily outweigh the expense of the die, and the package design typically plays a fundamental role in the function and performance of MEMS. MEMS packages must protect the sensor from the environment while enabling the device to perform its function. In some applications the package must protect the environment from the device (e.g., bioMEMS) and provide electrical connection. Pressure sensor packages must in some cases protect the sensor from the pressurized media while transmitting the pressure to the sensor. This is typically done by using a stainless steel barrier diaphragm and back filling the sensor cavity with a hydraulic fluid. Even basic packages must at least provide a pressure port and some isolation from unwanted mechanical stress. MEMS can employ similar packages to those developed for ICs (plastic, ceramic, and metal), but the fundamental interaction of the system with its surrounding invariably complicates the package design, meaning bespoke packages are often required for each application. Wafer level packaging using wafer bonding can simplify the final package design. For example, bonding a capping wafer over a wafer containing silicon accelerometers allows the individual die to be packaged in low-cost plastic packages. High-performance sensors will require carefully designed packages that provide stress relief from undesirable forces. Again, wafer bonding can be employed to fabricate, at wafer level, some first-order strain relieving packaging. Stress relieving glass intermediates and micromachined silicon mounts, for example, can isolate the sensor die from the package. Flexible adhesive can also be used to provide some stress relief.

2.2.2. Piezoresistive

Piezoresistivity derives its name from the Greek word *piezin*, meaning "to press." First discovered by Lord Kelvin in 1856, it is an effect occurring in various materials that exhibit a change in resistivity due to an applied force. The sensitivity of a strain gauge is defined by its gauge factor, which is a dimensionless quantity and is given by equation 2.10, where R is the initial resistance of the strain gauge and ΔR is the change in resistance.

$$GF = \frac{\text{relative change in resistance}}{\text{applied strain}} = \frac{\Delta R / R}{\Delta L / L} = \frac{\Delta R / R}{\xi} \qquad (2.10)$$

The term $\Delta L/L$ is the applied strain and is denoted as ξ (dimensionless). For all elastic materials, the stress σ (N/m²) and strain ξ obey Hooke's law and thus deform linearly with applied force. The constant of proportionality is the Young's modulus of the material given by equation 2.11:

$$\text{Young's modulus, } E = \frac{\text{Stress}}{\text{Strain}} = \frac{\sigma}{\xi} \quad (\text{N/m}^2) \qquad (2.11)$$

Silicon has anisotropic material properties (i.e., they vary with crystalline direction), but Young's modulus can be taken to be 190 GPa (1 Pa = 1 N/m²), which is close to that of a typical stainless steel (around 200 GPa).

The physics of a strain gauge are as follows. When an elastic material is subjected to a force along its axis, it will stretch but will also deform along the orthogonal axes. If, for example, a rectangular block of material is stretched along its length, its width and thickness will decrease. Typically, the magnitude of the axial and transverse strains will differ, and the ratio between the two is known as *Poisson's ratio, ν*. Most elastic materials have a Poisson's ratio of around 0.3 (silicon is 0.22).

Basic analysis of a resistive material under strain yields the following equation for the gauge factor:

$$GF = \frac{dR/R}{\xi_l} = \frac{d\rho/\rho}{\xi_l} + (1+2\nu) \qquad (2.12)$$

This equation indicates clearly that there are two distinct effects that contribute to the gauge factor. The first term is the piezoresistive effect $((d\rho/\rho)/\xi_l)$, and the second is the geometric effect $(1 + 2\nu)$. Poisson's ratio is usually between 0.2 and 0.3, so the geometric effect typically equates to 1.4 and 1.6. Conventional strain gauges are typically made from a thin metal foil mounted on a backing film, which can be glued onto a surface. The change in resistance of these gauges arises purely from the geometric effect as the length of the conductive path is stretched and its cross-sectional area reduced.

Some materials, however, are inherently piezoresistive. The application of strain to these materials results in a change in the resistivity of the material itself. Strain gauges that exploit such materials are known as piezoresistors. These resistors have a much higher gauge factor than metallic strain gauges, which rely purely on the geometric effect. Semiconductor strain gauges possess a very high gauge factor with p-type silicon exhibiting a gauge factor up to +200 and n-type silicon a negative gauge factor down to −125. Polysilicon has a gauge factor of ±30 and thick-film printed piezoresistors approximately 10. While these gauge factors result in higher sensitivity devices, semiconductor strain gauges are also very sensitive to temperature. Temperature compensation methods must therefore be adopted when using semiconductor strain gauges.

The piezoresistive mechanism in silicon results from the effect of applied stress on the effective mobility of majority charge carriers. With p-type silicon, the mobility of holes decreases and therefore the resistivity increases (positive gauge factor). For n-type materials, the effective mobility of the electrons increases and hence the resistivity decreases with applied stress (negative gauge factor). The effect is anisotropic; that is, it also varies with the orientation of the resistor with respect to the crystalline planes. Ignoring the small geometric effect, the fractional change in resistivity in semiconductor strain gauges is given by this equation,

$$\frac{d\rho}{\rho} = \pi_l \sigma_l + \pi_t \sigma_t , \qquad (2.13)$$

where π_l and π_t are the longitudinal and transverse piezoresistive coefficients and σ_l and σ_t are the corresponding stresses. The longitudinal direction is defined as that parallel to the current flow in the piezoresistor, while the transverse is perpendicular to it. The two coefficients are dependent on the crystal orientation, dopant type (p- or n-type), dopant concentration, and temperature. The temperature coefficient of piezoresistivity is around 0.25%/°C in both directions. The application of silicon piezoresistors in silicon pressure sensors highlights many of these principles, and this is discussed further below.

Polysilicon and amorphous silicon are also piezoresistive, but because they are comprised of crystallites, the net result is the average over all orientations. The temperature coefficient of resistance (TCR), however, is significantly lower than that of single-crystal silicon and is generally less than 0.05%/°C. By carefully choosing the doping levels, it is possible to reduce the TCR further.

2.2.2.1. PIEZORESISTIVE PRESSURE SENSORS

This section introduces piezoresistive pressure sensors and describes some devices in detail. Automotive pressure sensors based upon this technology are discussed in section 4. The use of diffused or implanted silicon piezoresistors is a much applied technique for measuring the strain in a micromachined silicon diaphragm, and this was one of the earliest applications of MEMS technology. The earliest devices employed silicon strain gauges bonded to metal diaphragms [3]. This approach was superseded by the use of diaphragms micromachined into the silicon, first by mechanical spark erosion and followed by wet isotropic etching [4]. This was not a batch approach, however, and device costs were high. The development of wet anisotropic etching greatly improved the range of diaphragm geometries that can be achieved and provided a low-cost, batch-compatible process for the fabrication of MEMS pressure sensors. Developments in anodic and fusion bonding, ion-implanted strain gauges, and surface micromachining have also reduced the size and improved the performance of piezoresistive pressure sensors.

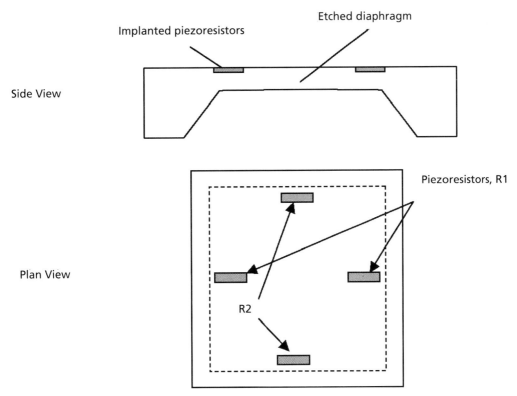

Figure 2.4. Anisotropically etched piezoresistive pressure sensor

A typical anisotropically etched silicon piezoresistive pressure sensor is shown in figure 2.4. The diaphragm is etched, and the resistors are located along the edge of the diaphragm, one on each side. The resistors are orientated in the same direction, two being parallel with the maximum strain (R1), two perpendicular (R2). Given a (100) orientation wafer, the edges of the diaphragm will be aligned in the (110) direction. Taking p-type doping, for example, which produces the largest and most linear piezoresistive effect, π_l and π_t are equal and opposite at typically ±69 m²/N. From equation 2.13, it can be seen that the resistor orientation shown in figure 2.4 produces equal and opposite changes in the resistance of R1 and R2. Placing the two pairs of resistors on opposite sides of a full Wheatstone bridge circuit will maximize the sensitivity to pressures applied to the diaphragm. This is the most common resistor arrangement and has been extensively modeled analytically [11,12].

Modifications to the basic diaphragm structure have been developed to improve the linearity and sensitivity of piezoresistive pressure sensors. Bossed diaphragms that incorporate a rigid center have been fabricated using anisotropic etching processes [6,13]. The boss is coupled with a resistor layout, as shown in figure 2.5, which enables equal and opposite strains to be experienced by the inner and outer resistor pairs. This arrangement improves the nonlinearity of the diaphragm in both directions, making it preferable for differential applications [14]. Another design uses a double boss at the diaphragm center [15], while researchers at Honeywell have used FE techniques to design a ribbed and bossed

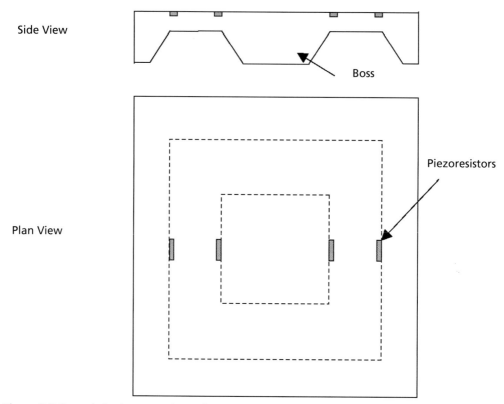

Figure 2.5. Bossed diaphragm resistor placement

diaphragm [16]. Meandering resistors, which incorporate different levels of doping in each direction, have also been applied to basic and bossed diaphragms [17]. The meander design maximizes the strain sensitivity of the resistor and increases the length of the resistor, thereby improving sensitivity.

The temperature cross-sensitivity is the main drawback associated with the use of silicon piezoresistors. The change in resistance due to temperature can exceed that arising from the change in pressure. Temperature compensation can be achieved by using a full bridge with the resistors arranged as shown in figure 2.4. In this arrangement the change in temperature is a common mode effect acting on all resistors, and therefore the temperature effect is largely cancelled out. However, manufacturing tolerances mean the temperature coefficients of each resistor will invariably be slightly different. Alternatively, temperature sensing can be incorporated onto the sensor chip, which enables temperature compensation via a lookup table or algorithm. This approach, however, requires extensive temperature and pressure calibration, which is time consuming and expensive. Another approach is to include a dummy bridge on the sensor chip with the dummy resistors positioned away from the edge of the diaphragm to ensure they do not experience any pressure-induced stresses [18]. The temperature limits of the implanted piezoresistive approach are approximately 125 °C due to the limitations of the p-n junction.

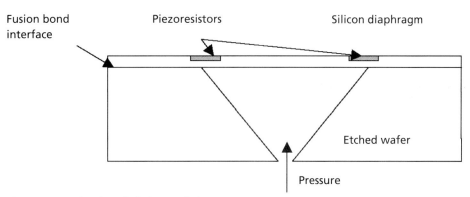

Figure 2.6. Fusion-bonded piezoresistive pressure sensor

Advances in materials and processes have affected the basic piezoresistive pressure sensor design. Silicon fusion bonding, for example, has enabled a reduction in chip size by enabling a diaphragm wafer to be bonded to the back of an anisotropically etched cavity, as shown in figure 2.6 [19]. Another example is the use of the buried oxide layer in SOI wafers as an etch stop provides a precisely controlled diaphragm thickness [20]. Silicon nitride diaphragms realized by bulk wet anisotropic etching through the silicon entirely from the back of the wafer have also been used. The wet etch stops upon reaching the nitride layer, and the piezoresistors are protected from the etch due to the high-dose boron implant used to define them, which renders them insoluble to the etchant [21]. Nitride membranes are stronger than their silicon counterpart but often suffer from built-in stresses from the deposition process.

Surface micromachining has also been used to fabricate piezoresistive pressure sensors [22]. Sensors with the diaphragm fabricated from polysilicon [23] and silicon nitride [24] have been described. In both cases an underlying sacrificial layer is removed, as shown in figure 2.3. Both devices use polysilicon resistors to sense diaphragm deflections and are used as absolute pressure sensors. A CVD process is used to deposit nitride and seal sacrificial etch holes, thereby trapping the process vacuum in the sealed volume under the diaphragm. A cross section of the polysilicon sensor is shown in figure 2.7.

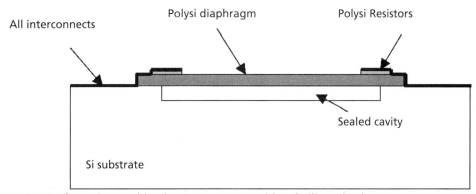

Figure 2.7. Surface micromachined pressure sensors with polysilicon diaphragm

2.2.3. Capacitive

Capacitive sensing provides a precise way of sensing movement and is a widely used sensing principle in MEMS. Capacitive sensors typically comprise a set of one (or more) fixed electrodes and one (or more) moving electrodes. The ability to integrate signal conditioning circuitry close to the surface micromachined capacitive sensors allows highly sensitive, temperature-compensated devices to be produced. Figure 2.8 illustrates three configurations for a simple parallel plate capacitor structure.

For a parallel plate capacitor structure, ignoring fringing fields, the capacitance is given by equation 2.14,

$$C = \frac{\varepsilon_0 \varepsilon_r A}{d} \quad \text{(Farads)}, \qquad (2.14)$$

where ε_0 is the permittivity of free space, ε_r is the relative permittivity of the dielectric material between the plates, A is the area of overlap between plates, and d is the gap. Capacitance can be varied by changing one or more of the variables, as shown in figure 2.8.

Capacitor structures are relatively straightforward to fabricate, and membrane-type pressure sensors and microphones are widespread. More elaborate structures, such as interdigitated capacitors, are also commonly used in MEMS.

Capacitive techniques are inherently lower noise than piezoresistance approaches due to the lack of thermal (Johnson) noise. In MEMS, however, the values of capacitance can be extremely small (often in the range femto to atto Farads), and the additional noise from the required interface electronic circuits can ultimately exceed that of a resistance-based system. There are a variety of approaches for measuring capacitance changes including charge amplifiers, charge balance techniques, AC bridge impedance measurements, and various oscillator configurations. A recent development has been to integrate planar coils on the capacitive pressure sensor chip. The capacitor and coil form a resonant LC circuit the frequency of which varies with applied pressure, and, by integrating the coil on the sensor chip itself, it can also be used to inductively couple power into the sensor chip from an external coil. The external coil can also be used as an antenna to detect the resonant frequency, and this approach is

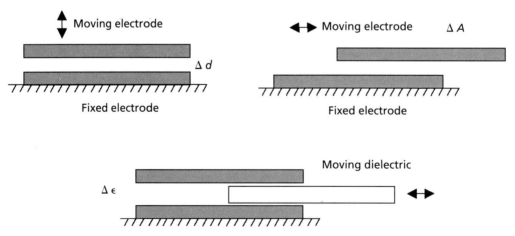

Figure 2.8. Examples of simple capacitance displacement sensors: (a) variable gap, (b) variable area, (c) moving dielectric

attractive for wireless sensing. Devices including electroplated coils [25,26] and a prototype microsystem on a ceramic substrate with a printed gold coil have been reported [27].

There are also a variety of commercially available capacitive interface ICs designed by a number of manufacturers such as Microsensors Capacitive Readout IC MS3110, Analogue Microelectronics CAV414, Xemics XE2004, and Smartec's Universal Transducer Interface chip. However, in order to reduce the effects of parasitic capacitance and achieve higher-performance devices, the pressure sensor should ideally be integrated with electronics. This typically requires a surface micromachining process and the combination of standard sacrificial surface micromachining processes with CMOS capacitance measurement circuitry [28]. The small size of these transducers can lead to a reduction in sensitivity, but this can be mitigated by simultaneously fabricating an array of diaphragms on the die. Using this approach, diaphragms with different pressure sensitivities have been incorporated onto the same die in order to broaden the range of operation [29,30].

Capacitive pressure sensors are typically based upon a parallel plate arrangement whereby one electrode is fixed and the other is able to deflect under applied pressure. The change in gap between electrodes alters the capacitance. An early capacitive pressure sensor is shown in figure 2.9. It consists of an anisotropically etched silicon diaphragm with the fixed electrode being provided by a metalized Pyrex 7740 glass die [31]. The glass and silicon dies are joined using anodic bonding. This device demonstrated the high sensitivity to pressure, low power consumption, and low temperature cross sensitivity achievable with capacitive pressure sensors.

The main drawbacks associated with the capacitive pressure sensing are the inherently nonlinear output of the sensor and the complexity of sensing electronics when compared with the resistive bridge (discussed above). The nonlinear output is inherent from equation 2.14, since the change in capacitance is inversely proportional to the gap height. In addition to this, a basic diaphragm will bend as it deflects and will therefore no longer be parallel to the fixed electrode, thereby introducing a further nonlinearity in the sensor output. The use of bossed diaphragms can reduce this effect [32,33]. Another linearizing approach is to sense the capacitance from a particular part of the diaphragm by patterning the electrodes. Maximum deflection occurs at the diaphragm center, but this is also the

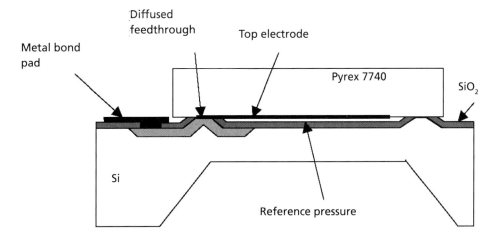

Figure 2.9. Early silicon/Pyrex capacitive pressure sensor

location of maximum nonlinearity. By sensing the capacitance at an annulus removed a short distance from the diaphragm center, nonlinearity is reduced but at the expense of reduced sensitivity [34,35]. Another approach, again at the expense of sensitivity, is to clamp the center of the diaphragm such that the pressure-sensitive structure becomes a ring shape. The sensitivity of such a structure is half that of a similar size standard flat-plate diaphragm, but nonlinearity falls to 0.7% FS [36]. Another approach is to operate the sensor in touch mode, where the diaphragm touches the fixed electrode. The center of the diaphragm is in contact with the fixed electrode, and the area in contact increases with increasing pressure [27,37,38]. Touch mode devices, however, suffer from hysteresis arising from friction between the surfaces as they move together and apart and the risk of stiction.

Similar devices to the surface micromachined pressure sensors have been realized using SOI wafers [39]. The buried oxide is used as the sacrificial layer, and a hole to allow the undercutting etch is located at the center of the diaphragm. The hole is sealed afterwards by silicon nitride deposition, which results in a ring-shaped diaphragm described above.

2.2.4. Resonant Pressure Sensors

At the heart of a resonant sensor is a mechanical structure designed to vibrate in a particular mode of vibration at a specific resonant frequency. Micromachining enables the resonant structure to be fabricated from a range of single-crystal materials with micron-sized dimensions. The resonant frequencies of single-crystal microresonators are extremely stable, enabling them to be used as a time base (e.g., the quartz tuning fork) or as the sensing element of a resonant sensor [40,41]. Resonant sensors provide a quasi-digital output, with much higher resolution and sensitivity than piezoresistive and capacitive techniques. However, the fabrication of such devices is more complex, and the requirement for packaging such devices more demanding.

A block diagram of a typical resonant sensor is shown in figure 2.10 [42]. A resonant pressure sensor is designed such that the resonator's natural frequency varies with changes in pressure. This requires the pressure to be coupled to the resonator in some manner, for example, by altering its stiffness, mass, or shape, and hence causing a change in its resonant frequency. The other components of a resonant sensor are the vibration drive and detection mechanisms. The drive mechanism excites the vibrations in the structure while the detection mechanism "picks up" these vibrations. The frequency of the detected vibration forms the output of the sensor. This signal can also be fed back to the drive mechanism via an amplifier maintaining the structure at resonance over the entire measurand range.

In mechanical sensing applications such as pressure, the most common mechanism for coupling the resonator to the measurand is to apply a strain across the structure. Stretching the resonator alters the effective stiffness of the structure. When used in this configuration, the resonator effectively becomes a resonant strain gauge. Coupling to the pressure is typically achieved by mounting the resonator in a suitable location on a pressure-sensitive diaphragm.

Resonant structures fabricated from GaAs and quartz materials can be excited and their vibrations detected by exploiting the piezoelectric nature of the materials [43]. This requires the deposition and patterning of metallic electrode materials on the surface of the resonator. The location and geometry of the electrodes should be carefully designed to maximize the electrical to mechanical coupling. Maximizing this coupling preferentially excites the desired resonant mode and maximizes the corresponding vibration detection signal.

Silicon is not intrinsically piezoelectric, and therefore other mechanisms must be fabricated on or adjacent to the resonator structure to excite and detect vibrations. There are many suitable mechanisms,

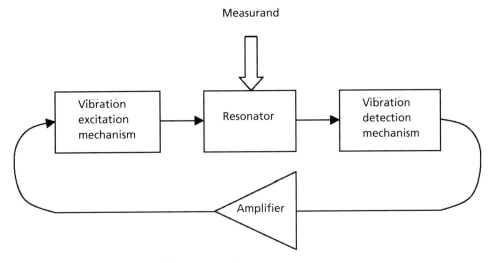

Figure 2.10. Block diagram of a silicon resonant sensor

and these are often used as sensors or actuators in their own right. For example, the resonator vibrations can be electrostatically excited and detected using implanted piezoresistors. The added complexity of the resonant approach is compensated by the increased performance.

As a structure being excited by a varying sinusoidal force approaches resonance, the amplitude of its vibration will increase. Its resonant frequency is defined as the point of maximum amplitude. The magnitude of the amplitude at resonance is limited by the damping effects acting on the system. The level of damping present in a system can be defined by its quality factor (Q factor). The Q factor is a ratio of the total energy stored in the system to the energy lost per cycle due to the damping effects present.

A high Q factor indicates a pronounced resonance easily distinguishable from nonresonant vibrations, and this will improve the performance and resolution of the resonator. It also improves the magnitude of the signal from the vibration detection mechanism, therefore simplifying the operating electronics. A high Q also means little energy is required to maintain the resonator at constant amplitude and implies the resonant structure is well isolated from its surroundings, and therefore the influence of external factors, for example, vibrations, will be minimized. The Q factor can be calculated by the following equation.

$$Q = \frac{f_0}{\Delta f_{3dB}} \qquad (2.15)$$

In this equation, resonant frequency f_o corresponds with a_{max}, the maximum amplitude, and f is the difference between frequencies f_1 and f_2. Frequencies f_1 and f_2 correspond to amplitudes of vibration 3dB lower than a_{max}.

The Q factor is limited by three types of damping mechanisms:

1. The energy lost to a surrounding fluid ($1/Q_a$)
2. The energy coupled through the resonator's supports to a surrounding solid (structural damping, $1/Q_s$)

3. The energy dissipated internally within the resonator's material ($1/Q_i$)

Minimizing these effects will maximize the Q factor as shown in equation 2.16.

$$\frac{1}{Q} = \frac{1}{Q_a} + \frac{1}{Q_s} + \frac{1}{Q_i} \qquad (2.16)$$

Energy lost to the surrounding fluid is the most significant. There are several distinguishable loss mechanisms (e.g., acoustic radiation, viscous drag, and squeeze film damping); and the magnitude of each depends upon the nature of the gas, the surrounding gas pressure, the size and shape of the resonator, the direction of its vibrations, and its proximity to adjacent surfaces. These damping mechanisms can be avoided by operating the resonator in a vacuum, and this is essential for micromechanical resonator applications. Structural damping ($1/Q_s$) is associated with the energy coupled from the resonator through its supports to the surrounding structure and is minimized by designing a balanced resonant structure, supporting the resonator at its nodes, or employing a decoupling system between the resonator and its support. For example, balanced resonator designs operate on the principle of providing the reaction to the structure's vibrations within the resonator. Multiple beam-style resonators (tuning forks, for example) incorporate this inherent dynamic moment cancellation when operated in a balanced mode of vibration. Examples of these are shown in section 2.3.3. Structural damping also provides a key determinant of resonator performance. A dynamically balanced resonator design that minimizes $1/Q_s$ provides many benefits [44]:

- High resonator Q factor and therefore good resolution of frequency
- A high degree of immunity to environmental vibrations
- Immunity to interference from and reduced chance of exciting surrounding structural resonances
- Improved long-term stability

The Q factor of a resonator is ultimately limited by the damping that occurs within the resonator material. This is illustrated by the fact that even if the other damping mechanisms are removed, the amplitude of its vibrations will still decay with time. Single-crystal materials such as silicon and quartz possess inherently low internal damping mechanisms and are therefore ideal resonator materials.

Resonant structures can exhibit nonlinear behavior, which becomes apparent at higher vibration amplitudes when the resonator's restoring force becomes a nonlinear function of its displacement. The nature of the effect and its magnitude depends upon the geometry of the resonator and the rigidity of its attachment and the surrounding structure.

The amplitude of vibration is dependent upon the energy supplied by the resonator's excitation mechanism and the Q factor of the resonator. Driving the resonator too hard or a Q factor that results in excessive amplitudes at minimum practical drive levels can result in undesirable nonlinear behavior. Taken to an extreme, a nonlinear system can exhibit hysteresis if the amplitude of vibration increases beyond a critical value. Hysteresis occurs when the amplitude has three possible values at a given frequency.

The technical challenges associated with resonant pressure sensors are as follows:

- Incorporating a mechanical resonator structure on top of a diaphragm
- In the case of silicon resonators, fabricating vibration excitation and detection mechanisms
- Vacuum encapsulation of the resonator to remove gas damping effects

The earliest micromachined resonant pressure sensor was developed by Greenwood [45] and later commercialized by Druck [46]. A butterfly-shape resonator is attached via four arms to pillars, which form part of the diaphragm (see figure 2.11). As the diaphragm deflects upwards under pressure, the angle on the arms causes the resonator to be tensioned and the resonant frequency to increase. The two halves of the resonator are coupled together via a small link, and the arms are positioned at node points in the optimum mode of operation. The resonator is driven electrostatically, and its vibrations detected capacitively. A vacuum is trapped around the resonator by mounting the sensor on a glass stem and sealing the end of the stem while in a vacuum. The resonator has a Q factor of 40000, the sensor has a resolution of 10 ppm and total error of less than 100 ppm [47].

Another successfully commercialized device has been developed by the Yokogawa Electric Corporation. The DPharp, EJA series differential pressure sensor [48] consists of two resonators located on a diaphragm, the differential output of which provides the sensor reading [49]. The resonators are driven electromagnetically by placing the device in a magnetic field and running an alternating current through the structure. The beams are vacuum encapsulated at wafer level using a combination of epitaxial depositions, selective etching, and an annealing step in nitrogen that drives the trapped gases left by the sealing process through the cavity walls or into the silicon. This leaves a final cavity pressure of below 1 mTorr and a resonator Q factor of over 50000 [50].

Other diaphragm-based resonant pressure sensors have been fabricated using silicon fusion bonding [51], surface micromachining combined with bulk etched diaphragms [52], SOI wafer technology [53], and entirely surface micromachined sensors [54]. Surface micromachined devices typically use comb drive structures to excite and detect lateral resonances, but the polycrystalline materials used to fabricate the resonator are inferior in stability and hysteresis to single-crystal silicon.

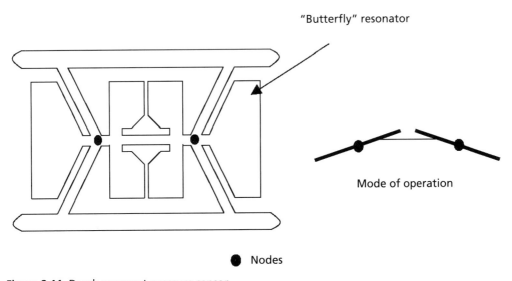

Figure 2.11. Druck resonant pressure sensor

2.3. MACHINED PIEZOELECTRIC CRYSTAL

2.3.1. Introduction

Piezoelectric crystal-based pressure sensors can either use the direct piezoelectric effect whereby a voltage is generated when a force is exerted on a piezoelectric material or the indirect piezoelectric effect whereby the application of a voltage produces a stress and a corresponding strain. There are a wide variety of piezoelectric materials available, naturally occurring quartz crystal and lead zirconate titanate being commonly used; others are lithium niobate, lithium tantalate, and barium titanate. There are three predominant approaches to utilizing the piezoelectric effect in pressure sensors: these being surface acoustic wave-based resonators, mechanically resonant, and those directly utilizing the piezoelectric effect. These are described in more detail in the following three sections.

2.3.2. Surface Acoustic Wave (SAW) Resonators

The first surface acoustic wave devices were made in 1965 [55], although SAWs' existence was first demonstrated by Lord Rayleigh in 1885. SAWs have a longitudinal component and a vertical shear component and propagate at the surface of a material where they are generated using interdigitated transducers (IDTs) on the surface of a material.

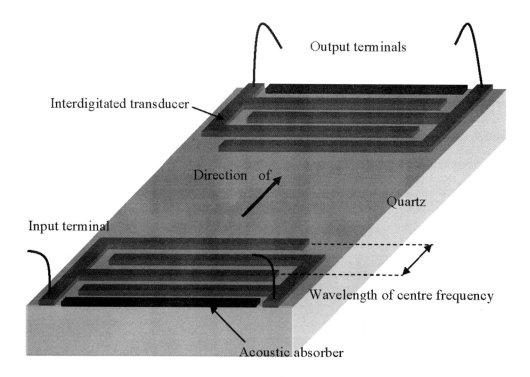

Figure 2.12. Basic surface acoustic wave delay line

If the material is piezoelectric, then the IDTs need only be metal films produced, for example, by evaporation. In this case there is an electric field associated with the propagating mechanical wave induced by the direct piezoelectric effect. If the substrate is not piezoelectric, SAWs can be produced by depositing piezoelectric film in an IDT structure by, for example, thick-film printing or sputtering. An alternating voltage applied to the IDT produced the SAW as a result of the indirect piezoelectric effect. SAWs have been launched on piezoelectric substrates such as quartz, lithium niobate, and gallium orthophosphate. In nonpiezoelectric substrates zinc oxide has been used as the piezoelectric material on silicon substrates, lead zirconate titanate has been used as the piezoelectric material on alumina/stainless steel substrates, and gallium nitride has been used on sapphire substrates.

The velocity of a SAW in a material is about 10^5 less than the corresponding electromagnetic wave of the same wavelength. The wavelength of the SAW is determined by the separation of the IDT fingers; for example, on quartz if the fingers are about 160 μm apart the SAW frequency is about 20 MHz.

A SAW is excited at one end of the substrate and detected at the other. A closed-loop system can be produced by connecting an amplifier between the launch IDT and the receive IDT. The oscillation frequency of the closed-loop system is modified by altering the time taken for the acoustic wave to travel from one IDT to the other. Clearly the separation of the IDTs can be altered by stressing the device, but in addition the velocity of the SAW is a strong function of stress [56]. A simple pressure transducer can be fabricated by incorporating the two IDTs on a diaphragm or with a bellows. As the diaphragm bends this changes the stress in the SAW propagation path and hence the SAW velocity and thus the frequency of the closed-loop system. Hydrostatic pressure can also be used to shorten the delay line path thus simplifying the sensors by avoiding the need for bellows or a diaphragm at the expense of reduced sensitivity.

SAW devices are suited to wireless, self-powered sensing applications since they can be remotely operated by radio-frequency electromagnetic waves. This approach is being explored commercially for tire pressure sensing applications and is described in more detail later in the section 4.10 on tire pressure sensing.

2.3.3. Quartz Resonant

Resonant quartz pressure sensors rely on the measurand of changing the properties of a vibrating quartz mechanical resonator [57]. Electrodes are deposited on the crystal. The direct piezoelectric effect is used to excite the vibrations of the resonator, and the indirect piezoelectric effect is used to detect the stress induced by the vibrations. The frequency of the mechanical resonator varies with stress applied to it, and this applied stress can be arranged to be a function of the pressure by using a force summing device.

Quartz is of interest in high-pressure measurement because of its high stiffness resulting in low mechanical hysteresis and its chemical inertness resulting in low corrosion. Applications containing fluoride acids or salts, however, will corrode quartz if in direct contact with the crystal. In this case an isolating diaphragm must be used between the quartz and the sensed media. In addition, the cut of the quartz crystal can be chosen to give a low or high temperature coefficient as required by the application.

Many novel designs exist for the mechanical resonator and its electrode structure. Important configurations based around flexural mode planar structures that can be easily integrated on a diaphragm are the single beam, the double-ended tuning fork [58], and the triple-beam tuning fork [59].

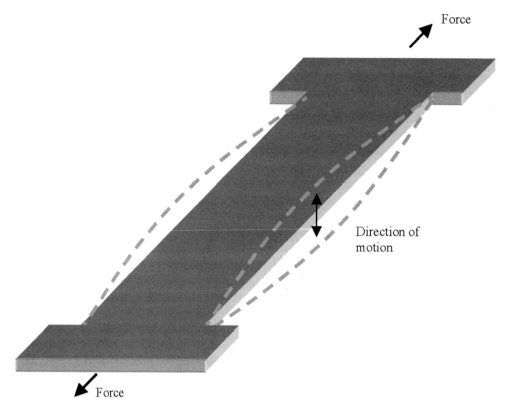

Figure 2.13. Single-beam flexural resonator

The multiple-beam designs attempt to circumvent the problem of mode coupling between the desired mode of the resonator and any resonant mode of the resonator's supporting structure that has a similar frequency. Should the desired mode of the resonator couple with another resonant mode, it will cause a reduction in quality factor and a slight pulling of the desired mode's resonant frequency. The double-ended tuning fork is based around the observation that the movement of a second beam can be used to cancel the moments and forces of the first beam. Such a concept is used in the well-known single-ended tuning fork to obtain a high-quality factor, and the double-ended tuning fork consists of two tuning forks joined together at their tine ends. When the beams move laterally in antiphase, a high Q factor mode is obtained; versions have been developed with and without a supporting stub and with and without outriggers [60]. Photolithographic machining of the structure allows the beam dimensions to be closely matched, thus resulting in a high Q. The triple-beam tuning fork allows a resonant mode out of the plane of the device rather than laterally. In this resonator the thickness of the tine controls its resonant frequency rather than the width, as in the single-beam and double-ended tuning fork. If the resonator is photolithographically defined, however, this width is controlled by the mask, whereas the thickness is more difficult to control and may require use of etch stop techniques such as those developed for silicon piezoresistive pressure sensor diaphragms. The coupling from a single-beam arrangement can be reduced by mechanically isolating the laterally flexing single beam by means of a

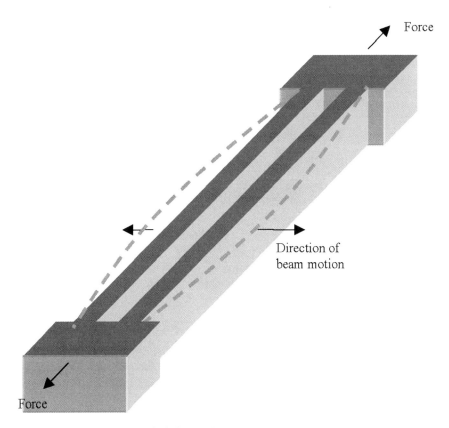

Figure 2.14. Double-ended tuning fork flexural resonator

spring mass system incorporated into both ends of the beam [61]. The spring is made flexible and the mass large, so the vibrational energy does not couple out of the vibrating beam and excite unwanted modes in the surrounding structure. This allows stresses to be applied to the beam but decouples the vibrating beam since the string mass combination acts as a low-pass filter for vibrations.

Other resonators investigated as sensors included tuning forks, torsional tuning forks, thickness shear, cylinders, hemispheres, and H- and star-shaped tuning forks. Resonating quartz diaphragms designed for lower pressures have been developed [62]. These have the problem that the density of the sensed medium will also change the frequency of the mechanical vibration.

Quartz is an attractive material for resonant applications given its piezoelectric properties and single-crystal material properties. The piezoelectric nature of quartz simplifies the excitation and detection of resonant modes, and quartz is routinely used in high-stability time-base applications. The main drawback associated with quartz is the limited choice of micromachining options compared with silicon and lack of suitability for integrating circuits within the quartz.

The corrosion of quartz in wet acids can be used as a machining technique to allow complex shapes to be defined in quartz wafers with accurate tolerances at small dimensions [63]. Etch rates depend on the crystallographic orientation with the slowest speeds corresponding to the densest

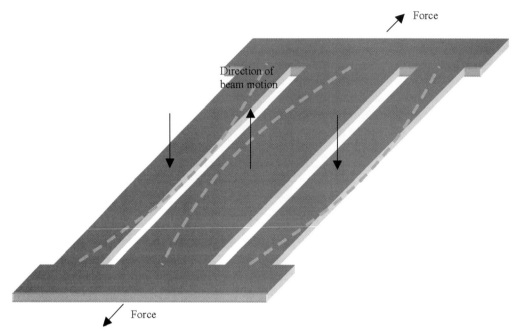

Figure 2.15. Triple-beam tuning fork flexural resonator

planes. Etchants used are hydrofluoric acid and ammonium fluoride with sputtered/evaporated chromium gold films as masks. Etch rates between 10 μmh^{-1} and 0.02 μmh^{-1} can be achieved depending on orientation. However, the range of structures micromachinable in quartz is highly limited when compared to silicon micromachining. Additionally, quartz lacks the range of dedicated supporting technology that silicon micromachining derives from the integrated circuit industry. For example, silicon micromachining can now produce high aspect ratio channels by virtue of recently developed deep reactive ion etching machines. No similar specifically developed process exists for quartz. In the area of wafer bonding, quartz can be fused by glass fusion bonding, but this is limited by the need to keep the bond temperatures below transition temperature (573 °C). Quartz wafers have been joined using a field-assisted process similar to silicon anodic bonding; however, this relies on depositing silicon on one quartz wafer and Pyrex on the other wafer so is clearly complex.

2.3.4. Quartz Piezoelectric

Quartz piezoelectric pressure sensors rely on the measurand stressing a piezoelectric quartz crystal thereby providing a charge proportional to the measurand using the direct piezoelectric effect. However, the charge generated rapidly decays through self-discharge. Therefore, piezoelectric sensors utilizing the direct effect cannot be used for steady-state or slowly changing pressures but are useful for rapidly changing pressures.

2.4. SCREEN-PRINTED THICK FILM

2.4.1. Introduction

Screen printing has been used for many years in the patterning of artwork on t-shirts and mugs but has also been used, since the 1950s, to manufacture hybrid circuits. It involves the deposition of an ink through a patterned screen of woven material onto the substrate or material positioned below the screen. The woven screen is patterned such that the ink can pass through open areas of the screen while in other areas the mesh is filled with a layer of emulsion. Hence the open pattern on the screen is transferred to the substrate. In the case of graphics, the inks are distinguished by their color. In microelectronic applications, the inks have a much greater range of functionality, and the printing requirements (e.g., resolution) are much more stringent.

Hybrid circuits typically comprise monolithic ICs, surface mount passive components, and thick-film layers used as interconnects, dielectric isolation, ground planes, and resistors. The cermet inks used consist of fine powders of typically < 5μm particle size mixed with a glass frit and an organic vehicle that gives the ink its viscous nature. The properties of the particles define the characteristics of the film. For example, metallic particles are used to form conductive inks and carbon particles form resistive inks. After deposition, the inks are dried at around 150 °C to evaporate off the bulk of the organic vehicle. They are then fired at temperatures up to 900 °C, which melts the glass frit thus bonding the film to the substrate. The glass frits are typically devitrifying, meaning they do not melt again when exposed to subsequent firing cycles. Lower-temperature inks are also available whereby the glass frit is replaced by a polymer binder that can be fired at temperatures below 200 °C.

The screens use a stainless steel, nylon, or polyester mesh held in tension within a metal frame. The emulsion is photosensitive, allowing the pattern to be formed on the screen by selectively illuminating the mesh using a photolithographic process. The printing process typically takes place on an alumina or stainless steel substrate, although silicon has also been used with micromachining processes and active thick-film materials to fabricate MEMS. The printing process uses an automated printer that comprises a floodgate and rubber squeegee, which are drawn across the surface of the screen. The floodgate smears the ink across the screen in one direction, and the rubber squeegee is then drawn across the screen in the opposite direction. The gap between the screen and substrate is typically around 0.5 mm. As the squeegee moves across the screen, the downwards pressure forces the screen into contact with the substrate and forces the ink through the openings in the mesh. The screen snaps back to its original position after the squeegee passes, and the ink is left on the substrate in the desired pattern. Screen printing is used in the fabrication of piezoresistive pressure sensors discussed in the following section and in the deposition of glass frit layers for bonding/sealing operations in sensor packaging.

2.4.2. Piezoresistive

Resistive cermet inks were found to be piezoresistive in nature [64]. Piezoresistivity was discussed in section 2.2.2. The gauge factor of thick-film resistors is around ten, some five times greater than conventional strain gauges. Piezoresistors can be printed directly onto stainless steel and ceramic diaphragms and produce a change in resistance with applied pressure. The choice of stainless steel is limited by compatibility with the high-temperature firing process. Stainless steel types 430 and 316 are compatible but are not necessarily the optimum steel for the pressure sensing applications. Therefore thick-film piezoresistive pressure sensors typically employ ceramic diaphragms. The resistors can be printed directly in a full-bridge arrangement and can be laser trimmed to balance the bridge. While not as compact or as sensitive as silicon devices, thick-film piezoresistive pressure sensors are relatively straightforward and low cost to fabricate and offer a higher gauge factor than metal strain gauge-based sensors. The ceramic diaphragm is also robust and chemically resistant to a wide range of potentially corrosive pressurized media.

2.5. SPUTTERED OR EVAPORATED THIN FILM

2.5.1. Introduction

Thin-film technology uses a variety of deposition processes to deposit a range of films directly onto the surface of the substrate. Many of these processes are used in semiconductor manufacturing such as CVD (see section 2.2.1.2), evaporation, and sputtering. Materials including dielectrics (e.g., silicon dioxide, silicon nitride), metals, and amorphous polysilicon can be deposited using these processes. Of particular interest in pressure sensing are evaporated or sputtered thin metal films typically less than 1μm thick. Evaporation is achieved by placing the target substrate and metal source in a vacuum and heating the source such that the metal enters the gaseous phase and condenses upon the unheated substrate. Sputtering involves the acceleration of energized plasma ions towards a target composed of the desired coating material. The impact of the incoming ions causes atoms from the target to be knocked off the surface with enough energy to travel to, and bond with, the substrate. Both of these processes coat the entire substrate and must be combined with a photolithographic masking and etching process to achieve the desired pattern.

2.5.2. Resistive Thin-film Pressure Sensors

Thin-film strain gauges use evaporated or sputtered thin metallic films deposited and patterned directly onto the substrate. The metal gauge works only on the geometric effect, and the maximum gauge factor that can be achieved is approximately 2. The advantage of this approach compared to bonding foil gauges (see section 2.7.3) is the elimination of the adhesive bonding step. The deposition processes result in metal films that are molecularly bonded to the diaphragm and are therefore much more stable, with resistance values that drift less. The reliability of the sensor also improves considerably. The gauges can be evaporated directly onto ceramic and metallic diaphragms to form the pressure sensor, although metal substrates will require an isolating dielectric layer to be deposited first. The metallic film can be patterned in a wide range of geometries enabling a full-bridge arrangement to be included on the diaphragm. The thermal stability of this type of pressure sensor is good, and it is well suited to high-temperature applications (see section 4.11).

2.6. OPTICAL AND OPTICAL FIBER-BASED TECHNIQUES

2.6.1. Introduction

Optical pressure sensing relies on modulating the properties of an optical frequency electromagnetic wave; the measurand can directly modulate the properties of the electromagnetic wave, which may be in free space or guided within an optical fiber or integrated waveguide. In the case of pressure sensors that use optical interfacing, the pressure sensor interacts with the measurand. The pressure sensor then modulates a property of the optical signal in order to provide an indication of the measurand. In general, optical sensors have found limited uptake for automotive pressure sensing in production vehicles, since their high cost when compared with alternative electrical techniques renders them uneconomic except in engine development programs. The following sections outline the basics of optical pressure sensing.

2.6.2. Intensity

Intensity variations are simply detected, because all optical detectors directly respond by providing an electrical output proportional to the incident optical intensity. Therefore, the pressure sensor must be arranged to vary the intensity of an optical signal, which is then incident on a photodetector. The diaphragm of a pressure sensor can be polished to a mirrorlike finish and therefore can be arranged to reflect light back into an optical fiber, as shown in figure 2.16.

Pressure-induced displacements of the diaphragm along the axis of the optical fiber will produce changes in the level of the reflected light coupled back into the fiber and therefore the intensity signal received at the photodetector [65]. The optical source is typically a light-emitting diode, since a coherent source is not required for intensity-based sensors. Normally, multimode optical fiber would be used to allow as much light as possible to be coupled into the fiber. Alternative optical sources could be an incandescent lamp or a laser, although the latter may present problems with modal noise.

Intensity-based systems suffer from variations in received intensity caused by factors not related to the measurand. For example, the intensity of output of an optical source will vary with time and temperature. For this reason intensity-based sensors often measure the optical source intensity and use this as a reference to compare with the signal modulated by the pressure-sensitive diaphragm. This can be achieved by splitting off a portion of the light before it is incident on the diaphragm; this problem complicates intensity-based sensors. Additionally, variations in the sensitivity of the optical detector can also cause difficulties and complications.

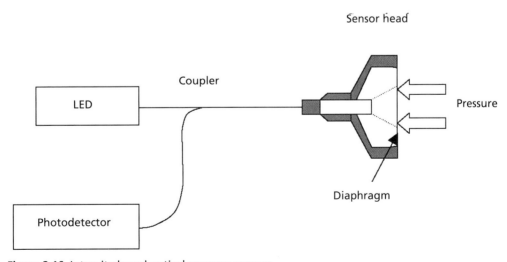

Figure 2.16. Intensity-based optical pressure sensors

2.6.3. Wavelength

Wavelength-based sensing relies on the source spectrum being modulated by interaction with the sensor. Normally, a source with a broad spectrum is used.

Figure 2.17 shows the incorporation of a Fabry-Perot cavity with a micromachined pressure sensor mounted on the end of an optical fiber [66]. Multiple reflections occur within the cavity, and a high intensity of reflected light occurs when there is constructive interference at the mirror nearest to the optical fiber. The condition for constructive interference is determined by the optical path length in the Fabry-Perot cavity, which is determined by the cavity spacing, the optical source wavelength, and the refractive index in the cavity. Assuming the refractive index is constant, the reflected wavelength of the constructively interfered light is therefore a function of the cavity spacing, which varies with pressure. An advantage of wavelength-based sensors is that they are insensitive to intensity variations, since these affect the whole spectrum in the same way. Photodetectors are not directly wavelength sensitive, so a prism in conjunction with a position-sensitive detector is often used as an analyzer.

2.6.4. Spatial Position

Figure 2.18 illustrates the principle of optical pressure measurement via the pressure-related modulation of spatial position. This technique is often known as triangulation.

Although resolution is lower than for phase-based techniques, the simplicity of optical pressure transducing based on movement of a "target" component, and its immunity to source intensity variations, makes this method suitable for automotive applications.

2.6.5. Phase

Photodetectors do not respond directly to variations in the optical phase, so it is necessary to convert phase variations to intensity variations for measurement by a photodiode. Often an interferometer is used to combine one or more optical beams that have interacted with the pressure with one or more optical beams that are unaffected by the pressure. A common configuration is to use an optical fiber Mach Zehnder interferometer with pressure exposed to one arm of the interferometer and the other arm used as a reference. The pressure may be directly exposed to the optical fiber that is relatively insensitive, or the fiber may be coupled to a diaphragm so it is strained as the diaphragm deflects.

Diaphragm-based pressure sensors have been fabricated with integrated optical waveguides on the top surface. Deflections in the diaphragm alter the phase of a light wave via the elasto-optic effect [67]. This is detected by having a reference waveguide unaffected by pressure and arranging the guides in a Mach Zehnder interferometer [68].

A major advantage of phase-based systems is that they are highly sensitive since subwavelength phase variations can be resolved, which equates to submicron displacements. Difficulties are caused by the periodic output of the interferometer; therefore, care has to be taken to establish the start pressure and subsequent relative pressure. This leads to complexity and errors in initializing the system.

2.7. CONVENTIONAL ASSEMBLY

2.7.1. Introduction

This section covers more traditional sensors, which were initially incorporated into vehicle control systems but have been, or are being, replaced by micromachined sensors, in particular those made from

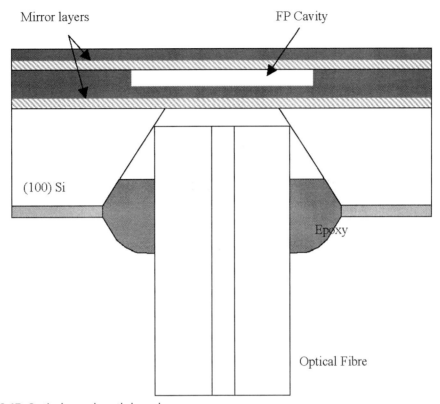

Figure 2.17. Optical wavelength-based pressure sensor

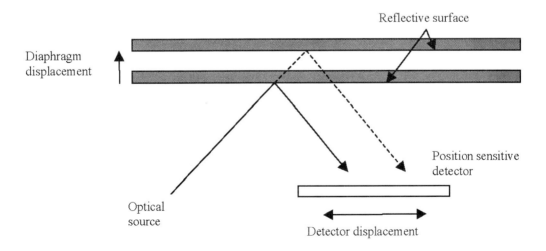

Figure 2.18. Spatial position measurement

silicon. Conventionally assembled sensors mean those that are typically not batch fabricated and which often require hand assembly processes to be performed on one sensor at a time. In general micromachined sensors have tended to be cheaper, smaller, and offer greater functionality when compared with their conventionally assembled counterparts.

2.7.2. Ceramic or Metal Capacitive

The basic principle of a capacitive pressure sensor has been discussed in section 2.2.3. They are typically fabricated from layers of ceramic or metal bonded together to form a sandwich structure that deforms under pressure, resulting in a change in capacitance. A conventional absolute pressure sensor uses a two-plate arrangement, similar to the micromachined device shown in figure 2.9. This is typically implemented with the reference vacuum forming the dielectric of the capacitor. This reference can then be sealed to prevent moisture or dirt ingress into the capacitor. Differential sensing can be achieved using a movable diaphragm between two fixed electrodes. If all three plates are isolated, this forms two capacitors that are equal at zero pressure corresponding to zero diaphragm deflection. Deflection of the diaphragm in response to pressure differentially changes the capacitances between the diaphragm and the metal plates. The capacitor may be arranged in an AC Wheatstone half bridge to null temperature effects and common mode signals. The three-plate arrangement is well suited to differential pressure measurement with the differential pressure applied across the diaphragm. Care must be taken in differential pressure measurement of gases to prevent the ingress of moisture between the plates owing to the high permittivity of water.

Diaphragm materials are metals, metal-coated glass, or ceramics. Stainless steels such as 174PH or 155PH are commonly used, but for highly corrosive environments inconel, hastelloy, or titanium are preferred. Ceramic diaphragms are sometimes used because of their chemical inertness, hardness, and superior corrosion resistance to metals.

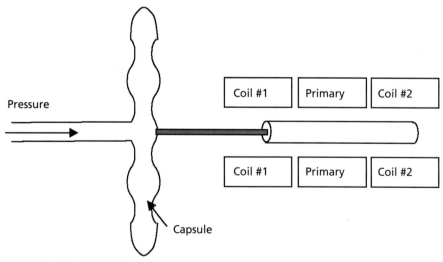

Figure 2.19. Principle of LVDT-based pressure sensor

2.7.3. Strain Gauge Bonded to Diaphragm

A simple form of pressure sensor uses a conventionally machined metal diaphragm onto which metal foil strain gauges are bonded. These can be produced at low cost. The foil gauges are bonded onto the back of the diaphragm so are protected from the sensed medium. The sensor body can be machined from the same single piece material as the diaphragm, avoiding the need to weld separate parts together, thus offering higher stability and reliability. Media compatibility is limited by the material of the diaphragm, so stainless steels and titanium are often used. The sensors are limited by the achievable location accuracy of the strain gauges and the sizes of diaphragms that can be machined and their tolerances. This technology is therefore more suited to higher pressures. The diaphragm area must be large enough to allow access to the diaphragm in order to attach the stain gauges, which are attached with adhesives. Since conventional foil gauges are used, temperature stability is good, but the strain sensitivity is low since there is no piezoresistive effect in metal gauges.

Some designs replace the foil gauges with bonded semiconductor strain gauges. These offer higher gauge factors than foil gauges at the expense of increased temperature coefficient and long-term drift. If a ceramic diaphragm is needed, gauges are more simply fabricated using thick-film technology as described earlier.

2.7.4. Linear Variable Displacement Transformer (LVDT)

In a linear variable displacement transducer (LVDT), three coils are wound on an insulating tube containing a ferrous core. The position of the core within the tube is controlled by the pressure. For example, it may be linked to a diaphragm or capsule arrangement by a connecting rod (as shown in figure 2.19). An alternating current is passed through the primary coil, which is the one in the center. If the core is in the center—that is, the diaphragm is undeflected at zero pressure—then equal voltages are induced in the secondary coils, these being the two outer coils.

The two outer coil outputs are differentially amplified, resulting in a zero output at zero pressure. As pressure changes a differential output from the coils is obtained from the coils according to the pressure that controls the position of the core. Accuracy of LVDT type pressure sensors is of the order of 0.5% FS.

2.7.5. Inductive Pressure Sensors

Inductive sensors use pressure-induced movement of a diaphragm to change the self-inductance of a single coil. Inductive pressure sensors operate on the principle that motion of a conductor in a magnetic field induces a voltage in the conductor. The inductive design is limited to dynamic measurements as the electrical output signal produced by the pressure requires relative motion. Inductive sensors require no external power supply, since they generate their own signal.

Inductive coupling has also been used on a MEMS pressure sensing by micromachining two planar coils, one fixed beneath a diaphragm and the other located on top of the diaphragm. An AC current is applied through the primary coil on the diaphragm, and the induced current in the second coil varies with applied pressure [69].

2.7.6. Reluctive Pressure Sensors

Reluctance is resistance to magnetic flow, that being the opposition offered by a magnetic material to magnetic flux. Reluctive sensors use pressure-induced movement to vary the magnetic coupling between a pair of coils; the magnetic coupling between the two coils is changed by pressure-induced movement of a conductor located in the magnetic field between the two coils. Reluctive sensors require AC excitation of one coil, and the signal is coupled to the second coil according to the position of the conductor. Reluctive signals typically have a very high output signal and are often used in applications where very high resolution is needed over a small range. Accuracy is of the order of 0.5% FS.

2.7.7. Potentiometric Pressure Sensor

A potentiometric pressure sensor uses a movable electrical contact, often called a wiper, attached so as to move in response to the pressure. The contact travels along a resistive element such as a conductive film or a wire wound coil. The resistive element is excited by AC voltage, and the voltage detected at the wiper is proportional to the position of the wiper on the resistive element and therefore the pressure. The two resistances produced between each end of the potentiometer and the wiper can be arranged in a Wheatstone half-bridge configuration to cancel out common mode effects such as temperature.

2.8. OTHER PRESSURE SENSOR PRINCIPLES

2.8.1. Introduction

This section simply covers other demonstrated pressure sensor principles, which may be of future interest in the field of automotive pressure sensors.

2.8.2. MOS Transistor

MOS transistors can also utilize the piezoresistive effect to sense strain and therefore pressure [70]. The piezoresistive effect alters channel carrier mobility and therefore the characteristics of the transistor [71]. In order to produce a sensor with a frequency-based output, the transistors are configured in a ring oscillator; this is achieved by fabricating an odd number of inverter elements using MOS transistors, which are then connected in a ring. The frequency of the ring oscillator is proportional to the number of gates and the average delay time per gate. The inverters are fabricated on a diaphragm, and their delay time varies with stress since the drain current varies with the mobility of the charge carriers.

2.8.3. Force/Pressure Balance

Force balance is an established sensing principle whereby an actuating force is applied to maintain the sensor structure in position during the application of the measurand. Electrostatic actuation has been applied to diaphragm structures for pressure sensing applications. The actuating voltage required provides a measure of the applied pressure [72]. This approach complicates the fabrication of the diaphragm, since an actuation electrode is required in addition to the diaphragm deflection sensing mechanism. However, this approach can improve dynamic range and linearity [73].

Another approach to achieve force balance of a microsensor uses the application of a restoring pressure on the other side of the diaphragm to the applied pressure [74] to keep the diaphragm movement within a limited range. A second high-accuracy pressure sensor measures the restoring pressure, and this value plus that from the microsensor gives the pressure reading. This approach allows the use of a very small microsensor in a confined location while giving a 24 dB improvement in dynamic range when compared to the same microsensor operated without pressure balance.

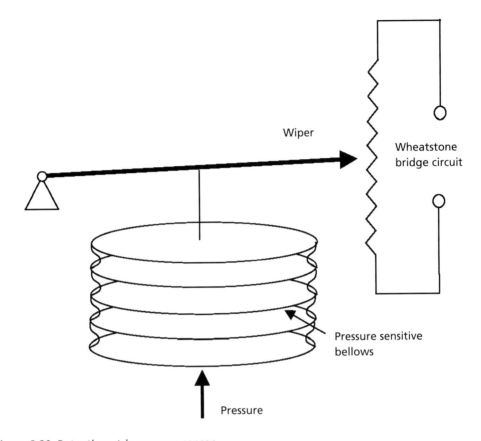

Figure 2.20. Potentiometric pressure sensor

3. HIGH-TEMPERATURE (> 125 °C) PRESSURE SENSORS SEGMENTED BY TECHNOLOGY

3.1. INTRODUCTION

This section gives an introduction to the approaches that may be adopted to achieve operation above a temperature of 125 °C in pressure sensors. A convenient temperature breakpoint of 125 °C is chosen since this is the limit of operation of silicon pressure sensors utilizing the piezoresistive effect. This limitation occurs since piezoresistors are formed by indiffusion of dopants into the silicon wafer. The piezoresistors are therefore isolated from the silicon wafer by a junction diode. As temperature increases, the reverse bias leakage current of junction increases, so that, above 125 °C, the piezoresistors can no longer be distinguished from the substrate resistance. This temperature limit can be extended by using doped polysilicon resistors deposited on the top surface of the diaphragm. Polysilicon resistors are, however, less sensitive to applied stress. Other approaches based on applying piezoresistors in high-temperature pressure sensors using silicon technology have therefore focused on dielectrically isolating the piezoresistors from the silicon substrate—silicon on insulator or silicon on sapphire technology.

Sensors of this type operate at least to 300 °C; limitations then arise from the metalization material and the means of electrically contacting the sensor.

3.2. SILICON ON INSULATOR

In this approach the silicon is used as a basic substrate into which the diaphragm is etched. The unetched surface of the wafer is coated with an insulating material; this is typically achieved by thermally oxidizing the wafer therefore producing a silicon dioxide coating. Piezoresistors are then produced in islands on the silicon dioxide layer by depositing polysilicon followed by metalization to connect them in a Wheatstone bridge configuration. Alternatively, silicon on insulator wafers are now commercially available in which a thick host wafer is bonded via a thin layer of silicon dioxide to a thinner silicon wafer. For example, for 100 mm diameter wafers, the thick wafer is around 525 μm thick, the insulating oxide 1 μm thick, and the thin wafer up to 10 μm thick. Alternatively, wafers can be fabricated with a buried oxide layer produced by oxygen implantation. The piezoresistors can then be photolithographically defined in the thin wafer and chemically or dry etched into separate islands before metalization [75,76]. This offers the advantage of the higher output and stability achievable from single-crystal piezoresistors compared with polycrystalline piezoresistors. Ultimate high-temperature piezoresistive pressure sensors have been realized using micromachined silicon carbide [77]. The diaphragms are etched by a photoelectrochemical process in a diluted HF etchant. A prototype device has been demonstrated operating at 600 °C [78] and in a dynamic sensing application on a gas turbine engine [79].

3.3. SILICON ON SAPPHIRE

Silicon on sapphire technology is very similar to silicon on insulator except in this case the diaphragm material is sapphire. Since sapphire is insulating, no oxide layer is required and single-crystal piezoresistors are directly produced on the sapphire by epitaxial growth. Sapphire also has very good elastic properties. Sensors of this type have been demonstrated to 425 °C [80]. Sapphire has the advantage of improved chemical resistance when compared to silicon and is harder; silicon also offers greater micromachining sophistication. Indeed, for sapphire, only direct mechanical machining methods can be used such as ultrasonic machining. A problem with this is mechanical machining can cause cracks in the sapphire, which lead to reduced sensor lifetime and performance. Therefore, to avoid machining the sapphire, it is often brazed directly to a titanium diaphragm. The titanium controls the pressure and corrosion characteristics, and the sapphire is host to the silicon piezoresistors. The thermal expansion coefficients of titanium and sapphire are well matched, and therefore thermal stresses and associated drift and reliability problems are reduced.

3.4. PIEZOELECTRIC MATERIALS

Alternative materials are inherently capable of high-temperature operation, for example, piezoelectric quartz undergoes a phase transition at 573 °C and so can be used up to this temperature. Lithium niobate can be used up to 600 °C for a limited time. Langasite has no Curie or phase transition up to its melting point of 1470 °C [81] and measurements of its resonant frequency have been obtained up to 700 °C. Gallium orthophosphate offers an operating temperature up to 900 °C. These materials can therefore be used as the basis of piezoelectric sensors such as those as described in section 2.3 while operating to higher temperatures. They have been utilized in engine development programs but have not been specified on production vehicles due to their prohibitively high cost. Recent interest has focused on the use of SAW resonators [82] to avoid the requirement to make electrical contact to the device at high temperatures; these are based on the approach described in section 4.10 on tire pressure sensors.

3.5. OPTICAL FIBER

It has been shown [83] that glass fibers suffer irreversible changes to their internal structures beyond temperatures of approximately 400 °C. Therefore, providing the fiber has a suitable cladding and terminations, optical sensors, such as those described in section 2.6, could be used up to this temperature. This assumes the sensor principle allows only the fiber to be exposed to the high temperature while the active components, e.g., optical source and detector, are maintained at temperatures < 125 °C.

Optical fibers can be produced in a wide range of single-crystal materials, and single-crystal sapphire fiber probes for temperature sensing have been demonstrated more than 20 years ago [84]. Devices for temperature measurement based on optical fibers are now available from Luxtron Corp. and can operate to 4000 °C. Therefore optical sensors should be capable of measuring pressure in an automotive environment; once again they have been utilized in engine development programs but have not been specified on production vehicles due to their prohibitively high cost. Another problem with optical fiber-based pressure sensing is that the fiber will often need to be interfaced to a mechanical structure such as the pressure-sensing diaphragm; the challenge is how to attach the diaphragm so that the structure first survives the elevated temperature and so the attachment does not introduce an excessive temperature coefficient.

4. AUTOMOTIVE APPLICATION AREAS TOGETHER WITH MOST APPROPRIATE TECHNOLOGY

4.1. INTRODUCTION

This section covers the actual technical approaches currently being investigated or currently in use in production vehicles. Inlet manifold pressure sensors are covered initially as this represents the most mature and well established of automotive pressure-sensing applications. Then, in turn, pressure sensors are covered for brake fluid, barometric pressure, exhaust gas, transmission, engine oil, and air conditioning. All these applications typically use an electrical piezoresistive pressure sensor with an isolation diaphragm if a corrosive medium is present. If extreme corrosion is present, a ceramic diaphragm-based pressure sensor is used. Lastly, newer sensor technologies for tire pressure monitoring and cylinder pressure are covered in more detail.

4.2. INLET MANIFOLD PRESSURE

Manifold absolute pressure sensors are employed in the air intake manifold to measure the engine load by detecting the sub pressure as well as the turbo charge pressure. The pressure measurement allows a calculation of air to fuel ratio. Early manifold absolute pressure sensors were based around linear variable displacement transducers measuring the displacement of a diaphragm. These have now been replaced by micromachined silicon sensors, which were introduced in 1979 and have been supplied in million-off quantities by the major market players. The advantages over competing technologies are smaller size and lower cost with similar performance.

Robert Bosch is an established supplier of such sensors, which are based on the piezoresistive effect. Other major suppliers are Delco and Ford. A diaphragm is etched out of a silicon wafer with the thickness being controlled by an electrochemical etch stop. This gives a more precise control of the diaphragm thickness than a simple timed etch. The diaphragm has four indiffused piezoresistors arranged in a Wheatstone bridge configuration. Pressure range is 4 bar absolute with an operating temperature range of –40 °C to 130 °C. Another example piezoresistive device has been developed by Motorola and is described in detail in Goldman 1998 [85].

An alternative approach, based on capacitance change, was used by Ford with the capacitor being formed between a bonded Pyrex wafer and a silicon wafer. Metallization on the Pyrex formed one plate of the capacitor, and the doped silicon wafer forms the other plate.

4.3. BRAKE FLUID PRESSURE

The 1995 S-class Mercedes was fitted with a piezoresistive pressure sensor in its master cylinder [86]. The sensors use a steel diaphragm on which polysilicon strain gauge elements are deposited. This avoids the need for silicone oil filling, which increases sensor cost and complexity and increases thermal lag within the sensor and hysteresis. The pressure range is 250 bar but the accuracy requirement is not stringent—of the order of 5% FS. The required temperature range is –40 °C to 120 °C. A typical sensor weight is of the order of 85×10^{-3} kg. Sensors must resist, in addition to brake fluid, mineral oils, water, and air. Robert Bosch supplies a device with an evaluation IC that can detect defects in bond wires, signal lines, or Wheatstone bridge supply, or ground. When it is switched on, the sensor runs a calibration and test routine.

4.4. BAROMETRIC PRESSURE

Barometric pressure measurement is needed to calculate the altitude of the vehicle to allow adjustment of the air to fuel ratio. These sensors are based on micromachined piezoresistive silicon and include an integrated evaluation circuit with a similar function to the type supplied by Robert Bosch with brake fluid pressure sensors. The measurement range is 0.5 bar to 1.1 bar and the sensors typically provide an analogue voltage output.

4.5. EXHAUST GAS RECIRCULATION PRESSURE

Conventional ceramic pressure sensors previously used in this application have been replaced by piezoresistive devices [87]. This is a highly corrosive environment so the sensors, which are now based on the piezoresistive principle, have an all stainless steel construction. They are guaranteed for 10 million cycles and have 0.15% FS stability over 1 year and an overall accuracy of 0.5% FS. Operating temperature is from –40 °C to 105 °C.

Czarnocki [88] has developed a differential pressure sensor based on two piezoresistive sensor exhaust gas recirculation systems. Even though this is a highly corrosive environment, the device does not need an isolation diaphragm or oil filling. The two corrosive pressure media are exposed only to the etched backside of the sensor, which does not contain any sensitive components. The front side, which contains the sensitive piezoresistors and interconnects, is exposed to the atmospheric pressure and forms a reference for each sensor. By appropriate signal condition a voltage signal can be extracted proportional to the difference in the two corrosive pressures.

4.6. ENGINE OIL/FUEL PRESSURE

Applications for oil or fuel pressure require operating temperatures between –40 °C to 130 °C at a low price including signal conditioning and are highly corrosive so need an isolation diaphragm. The technology is based around piezoresistive silicon strain gauges mounted on a stainless steel isolation diaphragm to cope with the corrosive environment. The required pressure range is 10 bar.

4.7. CONTINUOUSLY VARIABLE TRANSMISSION

The sensor needs to be immersed in hydraulic fluid, which, in the case of a silicon piezoresistive pressure sensor, means an isolation diaphragm must be incorporated.

4.8. AIR CONDITIONING

The requirement is to measure compressor pressure in the vehicle air conditioning system. Texas Instruments resistive ceramic pressure sensor is used, although this seems likely to be replaced by a micromachined solution. Other companies offering a micromachined alternative are Keller and Measurement Specialities.

4.9. COOLANT PRESSURE

Coolants are an extremely corrosive medium, and so ceramic capacitive pressure sensors have been employed primarily because of their high corrosion resistance. These have begun to be replaced by bulk micromachined silicon piezoresistive pressure sensors or silicon strain gauges. The silicon itself cannot be exposed to the sensor medium, so the devices are housed in sealed silicone oil–filled reservoirs and are coupled to a stainless steel or titanium diaphragm offering sufficient corrosion resistance in the environment.

4.10. TIRE PRESSURE SENSORS

Sensors for the measurement of pressure in vehicle tires have been under development for more than 15 years. Recently developments have become legislation driven, but early objectives were to increase safety and fuel economy through the use of tire pressure measurement. Very early systems used an electronic sensor circuit between slip rings or closely coupled coils in the wheel and a vehicle-mounted unit. Other systems made use of mechanical sensors in the wheel that conveyed a pressure threshold by means of a permanent magnet and a closely coupled coil. A third type tried to infer tire pressure from existing speed sensors but the accuracy of this approach was too low. Methods of sensor mounting used to date include the following:

- Within the tire on the rim using a stainless steel belt
- On the bottom end of the valve within the tire
- As a replacement valve cap

A useful review of the state of the art in tire pressure sensors was published as a result of a European Union–funded project [89].

Otter Controls in collaboration with ERA Technology and Centre Suisse Electronic et Microelectronique developed a sophisticated early tire pressure–sensing microsystem based around a silicon micromachined capacitive pressure sensor [90]. A schematic of the systems is shown in figure 2.21.

The microsystem also included a temperature sensor, a custom signal processing integrated circuit, a microprocessor, a motion sensor, a 433 MHz radio-frequency transmitter and an aerial. The whole system was powered by a lithium ion battery. Temperature and pressure measurements were made periodically at an update rate, typically 0.5 Hz, controlled by the microcontroller taking advice from the motion sensor. Following processing by the microcontroller, the data was transmitted to a central receiver unit incorporating an LCD display. The display allowed the driver to observe all tire pressures in real time and was powered from the car wiring system.

A micromachined silicon capacitive pressure sensor was developed because of its low power consumption when compared with silicon piezoresistive pressure sensors of a similar size and pressure range. Pressure sensor dimensions were 4 mm by 4 mm by 1 mm thick, and the sensor consisted of two conductive silicon wafers with an insulating layer of silicon dioxide between them, thus forming the capacitor. The operating range was 10 bar with a resolution of 0.06 bar, which allowed the same sensor to be used in both cars and commercial trucks.

In principle a capacitive pressure sensor could be operated with zero average power consumption, but in practice power is always consumed by the finite parallel resistance across the capacitor. This parallel resistance occurs, in the case of this structure, because of minute conductive particles produced during sensor dicing, which bridge across the silicon dioxide layer between the two conductive plates of the capacitor.

An application specific integrated circuit (ASIC) was designed to incorporate, on a single chip, the temperature sensor, signal processing IC, and the microprocessor. This allowed the total current consumption of the sensor module excluding the transmitter to be reduced to 20 µA from a 2.4V lithium ion battery (48 µW). For this application an important requirement is that the power consumption is maintained over a wide temperature range (–40 to 130 °C), and the system achieved a 5-year battery life over this range.

Wireless transmission was the largest overhead on the system and so strategies were developed to make transmission as infrequent as possible. To save power, data was only transmitted if the temperature or pressure reading had changed significantly, or if no new transmission had been made in the last 10 minutes when the vehicle was in use. If the vehicle was not in use the default transmission rate was 60 minutes. In this way a pragmatic approach to power saving was achieved while keeping the driver's display sufficiently updated.

Data transmission was via a 433 MHz link with data being replicated 4 times to ensure integrity. Included in the data transmission was a code uniquely identifying each tire. This code could be used

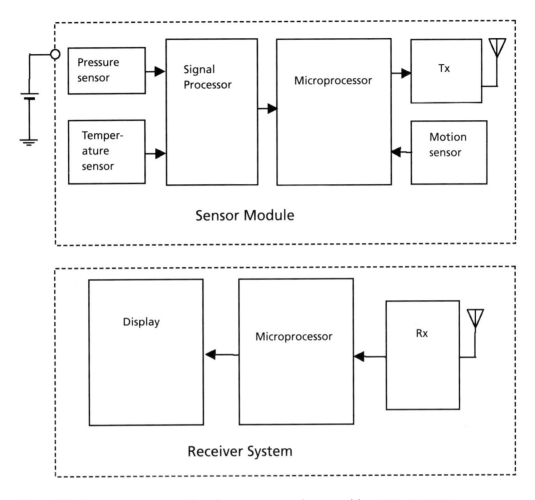

Figure 2.21. Low power consumption tire pressure–sensing capacitive pressure sensor

to avoid the receiver picking up erroneous tire pressure readings from an adjacent car and also could be used as identification in the event of wheel theft. Additionally, the transmission included battery state of charge data, allowing an indication of when the module's batteries should be changed.

Another early system, and one which has achieved considered commercial uptake, was developed by Schrader Electronics [91]. It has been fitted to vehicles from Ford, General Motors, Nissan, Daimler Chrysler, and Peugeot and is the market leader in remote tire pressure monitoring. From 2006 it was fitted on Mercedes cars. The system consists of four pressure sensors and transmitters in each wheel located on the valve and four receivers located in the wheel wells. Each pressure transmitter on the mark 2 version weighs only 34×10^{-3} kg including the valve and offers a 10-year battery life and 150000-mile durability. The driver may request an update of tire pressure as required.

Pressure can be measured up to 10 bar with a resolution of 0.013 bar with a ±2% of reading accuracy. Radio transmission frequency is selectable to suit the country requirements and super regenerative and super heterodyne AM/FM systems are available in 315, 433, or 868 MHz formats. The pressure sensors and transmitter system is rated from –40 °C to 125 °C. The central receiver operates from the 12 V battery supply and corrects for vehicle altitude-induced pressure changes. A mark 3 version will shortly be available, which reduces the weight of the transmitter module to 26×10^{-3} kg including the valve.

Another company offering battery-powered tire pressure sensors is SensoNor [92]. SensoNor offers a battery-powered pressure sensor, which communicates with an external microprocessor via an SPI interface. This minimizes power consumption at the sensor head but gives limited data analysis capability and a corresponding reduction in wireless transmission. As an alternative, a pressure sensor with an internal state machine has been developed that evaluates the measurement and presents it in a serial digital format. As is usual with micromachined pressure sensors, each device requires individual calibration to compensate for the effects of temperature. Following calibration, each sensor is paired with a signal conditioning ASIC and packaged together in a lead frame before encapsulation in epoxy resin.

Transense developed a tire pressure sensor system based on surface acoustic wave pressure sensing [93]. This technology has been licensed to tier one and two automotive suppliers of tire pressure systems. Licensees include SmarTire, 3DMI, and Michelin. The pressure sensor can be attached to the wheel, for example, on the back of the valve or even embedded directly in the car tire. The general principle of SAW-based tire pressure measurement is as follows [94]. A radio-frequency electromagnetic wave request signal is transmitted by a radar transceiver and picked up by the antenna of a passive SAW transponder. Connected to the antenna is an IDT, which converts the received signal into a SAW. The SAW propagates to a series of reflectors on the substrate, which are arranged in a bar code–like pattern, and is partially reflected off each one. The reflected SAWs are therefore returned to the IDT

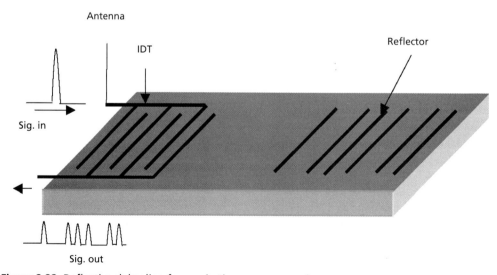

Figure 2.22. Reflective delay line for use in tire pressure sensing

and reconverted to electrical signal and then reradiated to the transceiver by means of the transponder antenna. A typical reflective SAW for this application is shown in figure 2.22.

The response contains information about the number and location of the reflectors as well as the propagation time of the SAW. The bar code arrangement can be used to uniquely identify each SAW sensor, and the delay time can be used to measure tire pressure. Thus the SAW sensor, being entirely passive, avoids the need for batteries within the tire and so greatly simplifies the pressure measurement at the expense of greater complexity at the transceiver. Care must be taken to ensure that environmental echoes caused by electromagnetic multipath propagation have faded by the time the sensor response arrives at the transceiver. This can be achieved by incorporating long delay times in the SAW device (of the order of microseconds); this can be achieved in a relatively small device by virtue of the relatively low velocity of SAW. Incorporating multiple SAW sensors at different crystallographic orientations within a single device allows both temperature and pressure to be measured independently.

4.11. CYLINDER PRESSURE

The cylinder pressure, by measuring directly the combustion process, is used to further optimize engine performance. This can increase fuel economy, reduce emissions, and improve engine performance. Sensors need to be embedded within the spark plug/ignition coils, fuel injectors, or glow plugs. Resistance to high temperature, small size, and immunity to electrical interference are therefore essential for this application. Additional requirements are long lifetime (10^9 cycles), low pressure hysteresis (< 0.2% FS), and high stability under high temperature (250 °C) and pressure cycling. Optical sensors are capable of withstanding the high temperature involved but until recently have been too expensive for widespread or even limited introduction. Piezoelectric pressure sensors have been used for many years by engine developers, but cost has prohibited their use in production vehicles [95]. Thin-film sensors, based on new materials, can address the required specification.

Optrand [96] via Honeywell commercially supplies a fiber optic–based pressure sensor with an adaptor to integrate it into the spark plug (see figure 2.23).

The sensor measures gauge pressure and is installed into the spark plug after fitting in the same way that the conventional spark plug connections are made. The sensor, which incorporates a light intensity referencing technique, uses one or multiple optical fibers positioned in front of a flexing metal diaphragm. The light intensity referencing technique compensates for optical fiber link fluctuations, diaphragm reflectivity variations, optical source intensity variations, and detector sensitivity variations. The sensor diaphragm communicates with the combustion cylinder via a pressure channel, which is formed in the threaded shell of the spark plug.

The intensity of the reflected light is proportional to the pressure-induced deflection of the diaphragm. To increase the modulation depth of the fiber-optic diaphragm sensor, the optical fiber(s) tip is either tapered or the diaphragm is mechanically leveraged relative to the tip end of the optical fiber. The diaphragm can be directly exposed to combustion temperatures in an internal combustion engine; the optical fibers not being in direct contact with the diaphragm are exposed to a lower temperature than that within the combustion cylinder.

Each optical combustion cylinder pressure sensor incorporates two optical fibers, an LED source, a photodiode, and a dedicated ASIC. The opto-electronic and ASIC components are capable of withstanding 150 °C with a total error of 1% of reading. The sensor head can be exposed to 300 °C. For commercial applications a lifetime of at least 500 million pressure cycles is required with low cost levels.

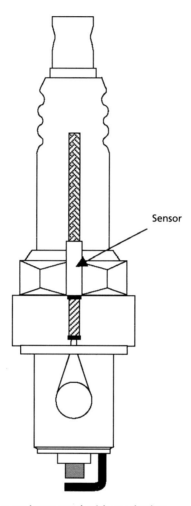

Figure 2.23. Optical pressure sensor integrated with spark plug

IMS GmbH supplies thin-film sensors for cylinder pressure measurement in diesel and gas engines [97]. Their performance compares with traditional pressure sensors offering low-temperature dependence and high linearity, and the sensors are capable of surviving 10^9 full-load cycles at temperature. The sensors are based upon strain gauges, which are evaporated or sputtered upon a steel diaphragm. This avoids the need for an oil separated diaphragm as is required in most piezoresistive micromachined silicon-based sensors used in vehicles. Initially silicon dioxide is plasma sputtered onto a polished surface steel diaphragm. Then layers of titanium oxy-nitride and nickel are precipitated and structured using a cathode atomizing process onto the silicon dioxide. The silicon dioxide provides the insulation, titanium the strain gauges, and the nickel the electrical interconnects. The layer's high strain sensitivity allows the use of thicker diaphragms and thus offers increased sensor longevity in a corrosive environment. This is followed by significant thermal aging at high temperature under stress to reduced drift.

4.12. DIESEL INJECTION PRESSURE

Diesel is injected into modern engines at pressures between 1300 and 2000 bar and such high levels require specialized high-pressure sensors [98]. Pressure sensors meeting this challenging specification have been developed that use an insulator material deposited onto the polished surface of a stainless steel diaphragm. Polysilicon piezoresistors or thin-film metallic resistors can then be deposited or sputtered onto the insulator and a four-bridge arrangement patterned and etched. Output is quite low at a few millivolts full scale due to the small size of the diaphragm, and this requires close proximity of amplifying and signal conditioning electronics. Other applications for such high-pressure sensors are ABS braking and petrol injection systems.

5. CONCLUSIONS

The primary technologies used for the fabrication of automotive pressure sensors and their main principles of operation have been reviewed. Automotive pressure sensors must be capable of reliable performance under challenging environmental conditions. Additionally the sensors must be mass produced in large volumes at low cost. These constraints suit sensor technologies that can be batch fabricated, and therefore micromachined silicon sensors have gradually replaced competing technologies. In some applications the silicon element itself can be directly exposed to the sensor medium (e.g., dry air), but in the majority of applications an intermediate diaphragm is placed between the silicon sensor chip and the environment. This allows the use of silicon sensor technology in highly corrosive environments since the diaphragm can be made of corrosion-resistant metals such as titanium. In the most corrosive applications, where such metals are not suitable, ceramic pressure sensors are used.

For high-temperature environments above 125 °C, silicon, optical, and thin film have been shown to be capable of addressing this extended temperature range. Cylinder pressure sensing requires extended temperatures but also subjects the sensor to high levels of wear. To date thin-film sensors have come closest to satisfying the requirements at the low cost levels demanded. Optical sensors are primarily of use in engine development programs, since they cannot to date satisfy the low cost requirements of production vehicles.

6. NOMENCLATURE

A is the area in m^2
c_o is the velocity of light in vacuum in $m\ s^{-1}$
C is the capacitance in F
°C is degrees centigrade
d is plate separation in m
kg is kilogram
v is velocity in $m\ s^{-1}$
V is volts
ε_0 is the permittivity of free space in $F\ m^{-1}$
ε_r is the relative permittivity of the dielectric in $F\ m^{-1}$
y_o is the maximum deflection at the diaphragm center
P is applied uniform pressure in Nm^{-2}

h is diaphragm thickness in m
a is the length of the shorter side in m
E is Young's modulus in Nm^{-2}
ν is Poisson's ratio
α and β are calculation coefficients for a rectangular diaphragm
y is diaphragm deflection in m
r is radial distance on a round diaphragm in m
a is the radius on a round diaphragm in m
f_n is the undamped resonant frequency of a clamped square diaphragm in radianss^{-1}
R is resistance in Ohms
ΔR is the change in resistance in Ohms
L is the length of a conductor in m
ΔL is the change in length of a conductor in m
ξ is strain ($\Delta L/L$)
σ is stress in Nm^{-2}
π_l is the longitudinal piezoresistive coefficients of a silicon piezoresistor
π_t is the transverse piezoresistive coefficients of a silicon piezoresistor
f_o is the resonant frequency in Hz
a_{max} is the maximum amplitude at resonance in m
f_1 and f_2 are the frequencies that correspond to amplitudes of vibration 3dB lower than a_{max} in Hz
Δf is the difference between f_1 and f_2 in Hz
Q_a is the damping arising from a surrounding fluid
Q_s is the damping arising from energy coupled through the resonator's supports to a surrounding solid
Q_i is the damping due to energy dissipated internally within the resonator's material

REFERENCES

1. Kruegar, S., R. Muller-Fiedler, S. Finkbieiner, and T. Hans Peter. 2005. Microsystems for the automotive industry. *MST News* S1/05: 8–10.
2. Tufte, O. N., P. W. Chapman, and D. Long. 1962. Silicon diffused element piezoresistive diaphragms. *Journal of Applied Physics* 33: 3322–27.
3. Sanchez, J. C. 1963. Semi-conductor strain gauge pressure sensors. *Instruments and Control Systems* (Nov.): 117–20.
4. Peake, E. R., A. R. Zias, and J. V. Egan. 1969. Solid-state digital pressure devices. *IEEE Transactions on Electron Devices* ED 16(19): 870–76.
5. Timoshenko, S. P. and S. Woinowski-Krieger. 1983. *Theory of plates and shells*. London: McGraw Hill.
6. Mallon Jr., J. R., F. Pourahmadi, K. Petersen, P. Barth, T. Vermeulen, and J. Bryzek. 1990. Low-pressure sensors employing bossed diaphragms and precision etch-stopping. *Sensors and Actuators* A21–A23: 89–95.
7. Sandmaier, H. 1991. Non-linear analytical modelling of bossed diaphragms for pressure sensors. *Sensors and Actuators* A25–27: 815–19.
8. van Mullem, C. J., K. J. Gabriel, and H. Fujita. 1991. Large deflection performance of surface micromachined corrugated diaphragms. In *Proceedings of Transducers '91, the 1991 International Conference on Solid State Sensors and Actuators*, 1014–17. Piscataway, NJ: IEEE.
9. Jerman, J. H. 1990. The fabrication and use of micromachined corrugated silicon diaphragms. *Sensors and Actuators* A23: 988–92.

10. Petersen, K. 1982. Silicon as a mechanical material. *Proceedings of the IEEE* 70(5): 420–57.
11. Clark, S. K., and K. D. Wise. 1979. Pressure sensitivity in anisotropically etched thin-diaphragm pressure sensors. *IEEE Transactions on Electron Devices* ED 26: 1887–96.
12. Elgamel, H. E. 1995. Closed-form expressions for the relation ships between stress, diaphragm deflection, and resistance change with pressure in silicon piezoresistive pressure sensors. *Sensors and Actuators* A50: 17–22.
13. Sandmaier, H., and K. Kuhl. 1993. A square-diaphragm piezoresistive pressure sensor with a rectangular central boss for low pressure ranges. *IEEE Transactions on Electron Devices* 40(10): 1754–59.
14. Kanda, Y., and A. Yasukawa. 1997. Optimum design considerations for silicon piezoresistive pressure sensors. *Sensors and Actuators* A62: 539–42.
15. Wu, X. P. 1990. A new pressure sensor with inner compensation for nonlinearity and protection to overpressure. *Sensors and Actuators* 21(1–3): 65–69.
16. Johnson, R. H., S. Karbassi, U. Sridhar, and B. Speldrich. 1992. A high-sensitivity ribbed and bossed pressure transducer. *Sensors and Actuators* A35: 93–99.
17. Dziuban, J., A. Górecka-Drzazga, U. Lipowicz, J. Indyka, and W. Wasowski. 1994. Self compensating piezoresistive pressure sensor. *Sensors and Actuators* A41–42: 368–74.
18. Suzuki, K. 1987. Nonlinear analysis of a CMOS integrated silicon pressure sensor. *IEEE Transactions on Electron Devices* ED 34: 1360–67.
19. Bryzek, J., K. Petersen, J. R. Mallon, L. Christel, and F. Pourahmadi. 1990. *Silicon sensors and microstructures*. Fremont, CA: Novasensor.
20. Diem, B., P. Rey, S. Renard, S. V. Bosson, H. M. F. Bono, M. T. Delaye, and G. Delapierre. 1995. SOI "SIMOX"; from bulk to surface micromachining: A new age for silicon sensors and actuators. *Sensors and Actuators* A46: 8–16.
21. Folkmer, B., P. Steiner, and W. Lang. 1996. A pressure sensor based on a nitride membrane using single-crystalline piezoresistors. *Sensors and Actuators* A54: 488–92.
22. Lin L., and W. Yun. 1998. Design, optimization and fabrication of surface micromachined pressure sensors. *Mechatronics* 8: 505–19.
23. Guckel, H. 1991. Silicon microsensors: construction, design and performance. *Microelectronic Engineering* 15: 387–98.
24. Sugiyama, S. 1986. Micro-diaphragm pressure sensor. In *Proceedings of IEEE IEDM International Electron Devices Meeting*, Los Angeles, CA, pages 184–87.
25. Akar, O., T. Akin, and K. Najafi. 2001. A wireless batch sealed absolute capacitive pressure sensor. *Sensors and Actuators* A95: 29–38.
26. deHennis, A., and K. D. Wise. 2002. A double sided single chip wireless pressure sensor. In *IEEE Tech Digest, 15th International Conference on MEMS*, 252–55. Piscataway, NJ: IEEE.
27. Suster, M., D. J. Darrin, and W. H. Ko. 2002. Micro-power wireless transmitter for high temperature MEMS sensing and communication applications. In *IEEE Tech Digest 15th International Conference on MEMS*, 641–44. Piscataway, NJ: IEEE.
28. Scheiter, T., H. Kapels, K-G. Oppermann, M. Steger, C. Hierold, W. M. Werner, and H-J. Timme. 1998. Full integration of a pressure sensor into a standard BiCMOS process. *Sensors and Actuators* A67: 211–14.
29. Dudaicevs, H., M. Kandler, Y. Manoli, W. Mokwa, and E. Spiegel. 1994. Surface micromachined pressure sensors with integrated CMOS read-out electronics. *Sensors and Actuators* A43(1–3): 157–63.
30. Chavan, A. V., and K. D. Wise. 2002. A monolithic fully-integrated vacuum sealed CMOS pressure sensor. *IEEE Transactions on Electron Devices* 49(1): 164–69.
31. Lee, Y. S., and K. D. Wise. 1982. A batch fabricated silicon capacitive pressure transducer with low temperature sensitivity. *IEEE Transactions on Electron Devices* ED 29(1): 42–48.
32. Zhang, Y., and K. D. Wise. 1994. An ultra sensitive capacitive pressure sensor with bossed dielectric diaphragm. *Technical Digest Solid-State Sensor and Actuator Workshop*, published by Hilton Head, SC, 205–8.
33. Beeby, S. P., M. Stuttle, and N. M. White. 2000. Design and fabrication of a low-cost microengineered silicon pressure sensor with linearised output. *IEE Proceedings of Science, Measurement and Technology* 147(3): 127–30.

34. Pons, P., G. Blasquez, and R. Behocaray. 1993. Feasibility of capacitive pressure sensors without compensation circuits. *Sensors and Actuators* A37–38: 112–15.
35. Hyeoncheol, K., Y-G. Jeong, and K. Chun. 1997. Improvement of the linearity of capacitive pressure sensor using an interdigitated electrode structure. *Sensors and Actuators* A62: 586–90.
36. Omi, T., K. Horibata, F. Satyo, and M. Takeuchi. 1997. Capacitive pressure sensor with centre clamped diaphragm. *IEICE Transactions on Electronics* E80-C(2): 263–68.
37. Park, J. S., and Y. B. Gianchandani. 1999. A low cost batch sealed capacitive pressure sensor. In *IEEE Tech Digest 12th International Conference on MEMS*, 82–87. Piscataway, NJ: IEEE
38. Wang, Q., and W. H. Ko. 1999. Modelling of touch mode capacitive sensors and diaphragms. *Sensors and Actuators* A75: 230–41.
39. Renard, S., C. Piscella, J. Collet, F. Perruchot, C. Kergueris, Ph. Destrez, P. Rey, N. Delorme, and E. Dallard. 2000. Miniature pressure acquisition microsystem for wireless in vivo measurements. In *Proceedings of 1st Annual International IEEE-EMBS Special Topic Conference on Microtechnologies in Medicine and Biology*, 175–79. Piscataway, NJ: IEEE.
40. Tudor, M. J., and S. P. Beeby. 1997. Resonant sensors: Fundamentals and state of the art. *Sensors and Materials* 9(3): 1–15.
41. Langdon, R. M. 1985. Resonator sensors—a review. *Journal of Physics E: Scientific Instrumentation* 18: 103–15.
42. Stemme, G. 1991. Resonant silicon sensors. *Journal of Micromechanics and Microengineering* 1: 113–25.
43. Eernisse, E. P., R. W. Ward, and R. B. Wiggins. 1988. Survey of quartz bulk resonator sensor technologies. *IEEE Transaction on Ultrasonics Ferroelectrics and Frequency Control* 35(3): 323–30.
44. Beeby, S. P., and M. J. Tudor. 1995. Modelling and optimisation of micromachined silicon resonators. *Journal of Micromechanics and Microengineering* 5: 103–5.
45. Greenwood, J. C. 1984. Etched silicon vibrating sensor. *Journal of Physics: Sci. Instrum.* 17: 650–52.
46. Druck. RPT (Resonant Pressure Transducer) Series Datasheet. http://www.druck-temperatur.de/index.html.
47. Greenwood, J., and T. Wray. 1993. High accuracy pressure measurement with a silicon resonant sensor. *Sensors and Actuators* A37–38: 82–85.
48. Harada, K., K. Ikeda, H. Kuwayama, and H. Murayama. 1999. Various applications of resonant pressure sensor chip based on 3-D micromachining. *Sensors and Actuators* A73: 261–66.
49. Ikeda, K., H. Kuwayama, T. Kobayashi, T. Watanabe, T. Nishikawa, T. Yoshida, and K. Harada. 1990. Silicon pressure sensor integrates resonant strain gauge on diaphragm. *Sensors and Actuators* A21–23: 146–50.
50. Ikeda, K., H. Kuwayama, T. Kobayashi, T. Watanabe, T. Nishikawa, T. Yoshida, and K. Harada. 1990. Three-dimensional micromachining of silicon pressure sensor integrating resonant strain gauge on diaphragm. *Sensors and Actuators* A21–23: 1007–10.
51. Petersen, K. 1991. Resonant beam pressure sensor fabricated with silicon fusion bonding. In *Proceedings of 6th International Conference on Solid State Sensors and Actuators (Transducers '91)*, 664–67. Piscataway, NJ: IEEE.
52. Welham, C. J., J. W. Gardner, and J. Greenwood. 1995. A laterally driven micromachined resonant pressure sensor. In *Proceedings of 8th International Conference on Solid State Sensors and Actuators (Transducers '95) and Eurosensors IX*, 586–89. Piscataway, NJ: IEEE.
53. Beeby, S. P., G. Ensell, B. Baker, M. J. Tudor, and N. M. White. Micromachined silicon resonant stain gauges fabricated using SOI wafer technology. *IEEE J. Microelectromechanical Systems* 9(1): 104–11.
54. Melvås, P., E. Kälvesten, and G. Stemme. 2001. A surface micromachined resonant beam pressure sensor. *IEEE Journal of Microelectromechanical Systems* 10(4): 498–502.
55. Morgan, D. 1998. History of SAW devices. In *Proceedings of IEEE International Frequency Control Symposium*, 439–60. Piscataway, NJ: IEEE.
56. Hauden, D., M. Planat, and J. Gagnepain. 1981. Nonlinear properties of surface acoustic waves: Applications to oscillators and sensors. *IEEE Transactions on Sonics and Ultrasonics* 5: 342–48.
57. Ward, R., and R. Wiggins. 1997. Resonant quartz pressure transducer technologies. http://www.quartzdyne.com.
58. Eernisse, E. 1980. Miniature quartz resonator force transducer. U.S. pat. 4215570.

59. Kirman, R., and S. Spencer. 1991. Vibrating force sensor. European pat. 0333377.
60. Clayton, L., S. Swanson, and E. Eernisse. 1987. Modifications of the double ended tuning fork geometry for reduced coupling to its surroundings: Finite element analysis and experiments. *IEEE Transactions on Ultrasonics, Ferroelectrics and Frequency Control* UFFC-34(2): 243–42.
61. Weisbord, L. 1970. Single tine digital force transducer. U.S. pat. 3505866.
62. Wagner, H., W. Hartig, and S. Büttgenbach. 1994. Design and fabrication of resonating AT-quartz diaphragms as pressure sensors. *Sensors and Actuators* A41–42: 389–93.
63. Danel, J., and G. Delapierre. 1991. Quartz: A material for microdevices. *Journal of Micromechanics and Microengineering* 1: 187–98.
64. Holmes, P. J. 1973. Changes in thick-film resistor values due to substrate flexure. *Microelectronics Reliability* 12: 395–96.
65. Economou, G., and D. Davies. 1987. Studies of an optical fibre displacement sensor. *J. Institution of Electronic and Radio Engineers* 57(2): 63–66.
66. Kim, Y., and D. Neikirk. 1995. Micromachined Fabry-Perot cavity pressure transducer. *IEEE Photonics Technology Letters* 7: 1471–73.
67. Yamada, A., Y. Shirai, T. Goto, M. Ohkawa, S. Sekine, and T. Sato. 2001. Relationship between sensitivity and waveguide position on diaphragm for silicon based integrated optic pressure sensor. In *IEEE Tech. Digest 4th Pacific Rim Conference on Lasers and Electro Optics*, I420–21. Piscataway, NJ: IEEE.
68. Benaissa, K., and A. Nathan. 1996. IC compatible optomechanical pressure sensors using Mach-Zehnder interferometry. *IEEE Transactions on Electron Devices* 43(9): 1571–82.
69. Okojie, R., and N. Carr. 1993. An inductively coupled high temperature silicon pressure sensor. In *Proceedings of 6th IOP Conference on Sensors and Their Applications*, 135–40.
70. Canali, C., F. Ferla, B. Morten, and A. Taroni. 1979. Piezoresistivity effects in MOS-FET useful for pressure transducers. *Journal of Physics D: Applied Physics* 12: 1973–83.
71. Alcántara, S. 1998. MOS transistor pressure sensor. In *Proceedings of IEEE International Conference on Devices, Circuits and Systems (ICCDCS '98)*, 381–85. Piscataway, NJ: IEEE.
72. Wang, Y., and M. Esashi. 1998. The structures for electrostatic servo capacitive vacuum sensors. *Sensors and Actuators* A66: 213–17.
73. Park, J., and Y. Gianchandani. 2003. A servo-controlled capacitive pressure sensor using a capped-cylinder structure microfabricated by a three-mask process. *Journal of Microelectromechanical Systems* 12(2): 209–20.
74. Hok, B., L. Tenerz, S. Berg, and A. Bluckert. 1994. Pressure microsensor system using a closed-loop configuration. *Sensors and Actuators* A41–42: 78–81.
75. Diem, B., R. Truche, S. Viollet-Bosson, and G. Delapierre. 1990. "SIMOX": A technology for high-temperature silicon sensors. *Sensors and Actuators* A23: 1003–6.
76. Kroetz, G. H., M. H. Eickhoff, and H. Moeller. 1999. Silicon compatible materials for harsh environment sensors. *Sensors and Actuators* A74: 182–89.
77. Okojie, R. S., A. A. Ned, and A. D. Kurtz. 1998. Operation of an (6H)-SiC pressure sensor at 500 °C. *Sensors and Actuators* A66: 200–204.
78. Ned, A. A., R. S. Okojie, and A. D. Kurtz. 1998. 6H-SiC pressure sensor operation at 600 °C. In *Proceedings of IEEE 4th International High Temperature Electronics Conference (HITEC)*, 257–60. Piscataway, NJ: IEEE.
79. Ned, A. A., F. Masheeb, A. D. Kurtz, and J. M. Wolff. 2001. Dynamic pressure measurements using silicon carbide transducers. In *Proceedings of 19th IEEE International Congress on Instrumentation in Aerospace Simulation Facilities*, 240–45. Piscataway, NJ: IEEE.
80. Keller, H., and A. Anagnostopoulis. 1987. Silicon on sapphire: the key technology for high-temperature piezoresistive pressure transducers. *Transducers '87*: 316–19.
81. Wolff, U., F. Dickert, G. Fischerauer, W. Greible, and C. Ruppel. 2001. SAW sensors for harsh environments. *IEEE Sensors* 1(1): 4–13.
82. Buff, W., M. Binhack, S. Klett, M. Hamsch, R. Hoffman, F. Krispel, and W. Wallnofer. 2003. SAW resonators at high temperatures. *IEEE Ultrasonics Symposium* 1: 187–91.
83. Zhang, Z., K. Grattan, and A. Palmer. 1992. Fiber optic temperature sensor based on the cross referencing between blackbody radiation and fluorescent lifetime. *Review of Scientific Instruments* 63: 3177–81.

84. Dils, R. 1983. High-temperature optical fiber thermometer. *Journal of Applied Physics* 54: 1198–2000.
85. Goldman, K., et al. 1998. A vertically integrated media-isolated absolute pressure sensor. *Sensors and Actuators* A66: 155–59.
86. Grace, R. 2000. Application opportunities of MEMS/MST in the automotive market: The great migration from electromechanical and discrete solutions. http://www.rgrace.com.
87. Kruegar, S., and R. Grace. 2001. New challenges for microsystems technology in automotive applications. *MST News* 1(1): 4–7.
88. Czarnocki, W. S. 1998. Media-isolated sensor. *Sensors and Actuators* A67: 142–45.
89. Appolo Consortium. 2003. Intelligent tire for accident-free traffic. http://www.vtt.fi/tuo/projects/apollo/index.htm.
90. Siddons, J., and A. Derbyshire. 1995. Smart low-power microsystems for automotive applications. In *Proceedings IoP conference "Sensor '95"*.
91. McClelland, S. 1996. Remote tire pressure monitoring system. U.S. pat. 5963128.
92. Grelland, R. 2001. Tyre pressure monitoring microsystems. *MST News* 1(1): 39–40.
93. Marsh, D. 2004. Tyre-pressure monitoring. *EDN Europe, 6-10-2004.* http://www.edn.com/article/CA421537.html.
94. Reindl, L., A. Pohl, G. Scholl, and R. Weigel. 2001. SAW-based radio sensor systems. *IEEE Sensors Journal* 1(1): 69–78.
95. Westbrook, M. 1989. Future developments in automotive sensors and their systems. *Journal of Physics E* 22(9): 693–99.
96. Jeschke, J. 2004. Conception and test of a cylinder pressure based engine management for passenger car diesel engines. PhD diss., University of Magdeburg.
97. Neumann, S. 2003. High temperature pressure sensors based on thin film technology for applications on combustion engines including modular electronic concept of data acquisition and processing. In *Proceedings of the 9th Symposium on the Working Process of the Combustion Engine*. University of Graz.
98. Marek, J., and M. Illing. 2002. Micromachined sensors for automotive applications. In *Proceedings of IEEE, Sensors 2002, 1st International Conference on Sensors 2*, 1561–64. Piscataway, NJ: IEEE

CHAPTER 3

TEMPERATURE SENSORS

John Turner

1. INTRODUCTION AND OVERVIEW

Temperature is an important parameter for automotive engineers. The first temperature sensors to be applied to motor vehicles were probably those based on bimetallic strips or mercury thermometers and used to monitor engine coolant (see figure 3.1). Thermistor-based temperature sensors have been used in automotive applications since the late 1940s to send signals to dashboard gauges, and later to electronic control modules. In modern vehicles temperature transducers are used in connection with electronic fuel control systems, and to measure the temperature of inlet air and exhaust gas. They are also used for environmental control, to regulate heating or air conditioning in the passenger compartment, and for ice warning systems. The majority of automotive temperature sensors are thermistors, though other forms of transducer are increasingly used.

The two primary users of temperature data on a vehicle are the engine/powertrain management system, and the heating, ventilation and air conditioning (HVAC) system. Additionally, many electronic modules require temperature compensation. This is normally provided by the use of a board-mounted thermistor.

The engine/powertrain management system uses a number of temperature inputs to enhance the performance of the engine, control emissions, and optimize efficiency. The most common applications are as follows:

- Coolant temperature sensing
- Intake air temperature sensing
- Transmission oil temperature sensing
- Cylinder head temperature sensing

The coolant temperature sensor measures the temperature of the coolant and interfaces with the electronic engine control module (ECM). This sensor provides feedback to the ECM regarding the

Figure 3.1. Early mechanical temperature transducer

temperature of the coolant at a single point on the engine. Similarly, the cylinder head temperature sensor provides the temperature of the metal at a single point on the engine. The ECM uses temperature measurement and previous engine calibrations to achieve optimal operation of the engine management system.

Engine intake air temperature is also used by the ECM. The intake air temperature sensor is normally located in one of three areas: within the air cleaner, in the intake air duct, or in the intake manifold. The intake air temperature transducer is often integrated with the mass airflow sensor or manifold absolute pressure sensor, to provide a multifunctional sensing unit.

In automatic transmissions a temperature sensor monitors the temperature of the automatic transmission fluid. This allows the powertrain control system to regulate the transmission's operation. It also maintains an optimal fluid temperature to reduce degradation and potential overheat conditions.

An emerging application for temperature sensing is the exhaust gas or catalytic converter sensor. This application is more demanding than the previously mentioned sensors due to the temperature range required. In many of these applications, the sensor is required to operate from –40 °C to 1000 °C.

Engine and powertrain temperature sensor applications are very common. However, the recent expansion of automatic HVAC control and driver information feedback has led to increasing applications for temperature sensors as part of the HVAC system. The most common applications for HVAC control are listed here:

- Outside air temperature sensing
- Cabin temperature sensing
- Duct air temperature sensing

The outside air temperature sensor monitors the temperature of the air external to the vehicle. This sensor supplies data to the HVAC control unit and/or display module to enable automatic HVAC control, or outside temperature display. The inside/cabin temperature sensor provides the temperature within the passenger compartment of the vehicle. Measurement of this temperature, and of course the outside air temperature, is essential if an automatically controlled cabin environment is to be provided. Duct sensors within the HVAC air delivery system are increasingly common to provide temperature "zoning" within the vehicle cabin. These sensors can further enhance the automotive HVAC system by providing additional measurement locations.

Temperature sensors can also be used for liquid level sensing, to control diesel fuel dewaxing, and for deicing systems such as those applied to heated exterior mirrors and windscreen washer jets.

Most of the electrical temperature sensors used in automotive engineering are either resistive or thermoelectric. Resistive devices may be either metallic or semiconductor, and require some form of bridge circuit for signal conditioning since they are modulating transducers. Thermoelectric sensors or thermo-couples are self-generating, but their very low output means that an amplifier is always needed in practice. As noted earlier, bimetallic temperature sensors were very common in the past and are still sometimes used. The thermal expansion of a solid is used as a primitive temperature transducer to control thermostat opening in water-cooled engines. Infrared emission or pyrometry has been proposed a number of times for automotive temperature sensing, and may be particularly useful for monitoring inaccessible or rotating components such as driveshaft joints (where the temperature of the joint bearings may give an early warning of failure). However, pyrometric sensors have yet to appear on a production vehicle.

Other temperature-based sensors that have yet to get beyond the research stage as far as automotive applications are concerned are heat flux gauges. These measure the rate at which heat is being transferred to or from a body, rather than its temperature. These devices have been included and are discussed in section 9, since it seems likely that they may find automotive applications in the near future.

2. RESISTIVE TEMPERATURE TRANSDUCERS

Resistive temperature transducers are the most common type for automotive use. They can be either metallic or semiconductor based, but usually are the latter since they are cheaper. They are sometimes known as *resistance temperature detectors* or RTDs. Metallic resistive temperature sensors offer better performance than semiconductor RTDs and may be preferred if high accuracy is required.

2.1. METALLIC RESISTIVE TEMPERATURE SENSORS

Metallic resistive transducers are similar in appearance to wire-wound resistors and often take the form of a noninductively wound coil of a suitable metal wire such as platinum, copper, or nickel. They may be encapsulated within a glass rod to form a temperature probe, which can be very small in size. An alternative form is shown in figure 3.2, in which a rectangular matrix of platinum is deposited on a ceramic substrate. Connections within the matrix are laser-trimmed to give a precise value of nominal resistance.

The variation of resistance R with temperature T for most metallic materials can be represented by an equation of the form
$$R = R_o\left(1 + a_1 T + a_2 T^2 + \ldots + a_n T^n\right),\tag{3.1}$$
where R_o is the resistance at temperature $T = 0$. The number of terms necessary in the summation depends on the material, the accuracy required, and the temperature range to be covered. Platinum, nickel, and copper are the most commonly used metals, and they generally require a summation containing at least two of the a_n constants for accurate representation. Tungsten and nickel alloys are also frequently used. In automotive applications it is often possible to model a metallic RTD using only constant a_1: with platinum, for example, $a_1 \approx 0.004$ (R in Ω and T in K). If this value is substituted into equation 3.1, and a_2, a_3, and so forth are set to zero, the resulting nonlinearity is only around 0.5% over the temperature range -40 to $+140$ °C.

The nominal resistance (R_o) of a metallic RTD can vary from a few ohms to several kilohms. However, 100 Ω is a fairly standard value. The resistance change of a metallic RTD can be quite large, and is typically up to 20% of the nominal resistance over the design temperature range.

2.2. THERMISTORS

Thermistors are small semiconducting transducers, usually manufactured in the shape of beads, disks, or rods. They are made by combining two or more metal oxides. If oxides of cobalt, copper, iron, magnesium, manganese, nickel, tin, titanium, vanadium, or zinc are used the resulting semiconductor has a negative temperature coefficient (NTC) of resistance. This means that as the temperature rises the electrical resistance of the device falls. Most thermistors used in automotive engineering are of this type, and they can exhibit large resistance variations. Typical values are 10 kΩ at 0 °C and 200 Ω at 100 °C. This very high sensitivity allows quite small temperature changes to be detected. However, the accuracy of a thermistor is not as good as that of a metallic RTD because of variations in the composition of the semiconductor that occur during manufacture. Most thermistors are manufactured

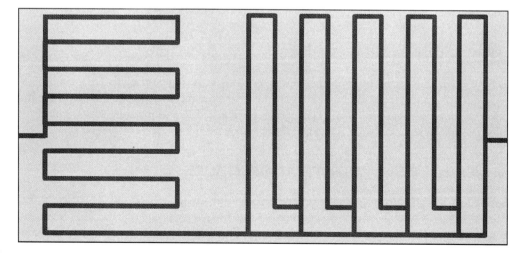

Figure 3.2. Miniature platinum resistance transducer

and sold with tolerances of 10 or 20%. Any circuit using semiconductor thermistors must therefore include some arrangement for trimming out the error.

Unlike metallic RTDs, thermistors are nonlinear. The resistance-temperature relation is usually of the form

$$R = R_0 \exp\beta\left(\frac{1}{T} - \frac{1}{T_0}\right), \qquad (3.2)$$

where R is the resistance (in ohms) at temperature T and the nominal resistance at a reference temperature T_o is R_o. The constant β is characteristic of the thermistor material and is typically around 4,000. T_o is usually taken to be 298 K (which is 25 °C).

Near to $T = T_o$ the curve of equation 3.2 becomes fairly linear, and a temperature coefficient of resistance α may be derived:

$$\frac{dR}{dT} = R\frac{d}{dT}(\beta\left\{\frac{1}{T} - \frac{1}{T_o}\right\}) = \frac{-R\beta}{T^2}$$

Defining $\alpha = \frac{1}{R}\frac{dR}{dT} = \frac{-\beta}{T^2}$, the equation becomes

$$\alpha = \frac{-\beta}{T_o^2} \text{ when close to } T = T_o. \qquad (3.3)$$

So, for $\beta = 4000$, $\alpha \approx -0.045 \Omega/°C$ at temperatures in the vicinity of 298 K. Using this coefficient,

$$R \approx R_0 (1 + \alpha \Delta T) \qquad (3.4)$$

when R is close to R_0 and ΔT is a small change in temperature.

Thermistors can be used within the temperature range from –60 to + 150 °C, which comfortably covers most of the temperature measurements required in automotive engineering. The accuracy can be as high as ±0.1%. The main problem associated with thermistors is their nonlinearity, as expressed by equation 3.3.

Positive temperature coefficient (PTC) thermistors can also be made using compounds of barium, lead. or strontium. PTC thermistors are usually only used to provide thermal protection for wound equipment such as transformers and motors. The characteristics of a PTC device have the form shown in figure 3.3. It can be seen that the resistance of a PTC thermistor is low (and reasonably constant) below the switching temperature T_R. Above this point the resistance rises spectacularly. In use PTC thermistors are often embedded in the windings of the equipment to be protected, and are connected in series with the power supply. If the temperature becomes too high, the resistance rises and power is effectively disconnected from the load. PTC thermistors are better than negative temperature coefficient (NTC) types for this sort of thermal protection task since they are fail-safe. If a connection to a PTC sensor fails, the resulting high impedance will disconnect the power. If the same happens to a thermal protection circuit containing an NTC thermistor, a false "low temperature" indication will be given, and full power will be applied.

"Conventional" thermistors are manufactured as discrete components. However, it is also possible to print thermistors onto a suitable substrate using thick-film fabrication techniques. Thick-film thermistors have very low cost, are physically small, and have the further advantage of being more intimately bonded to the substrate than a discrete component.

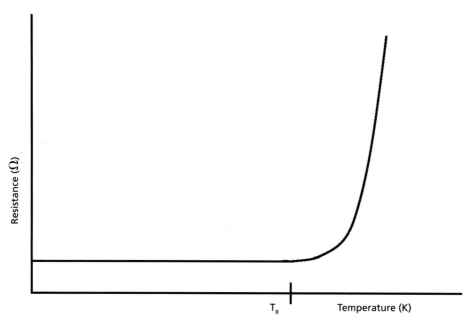

Figure 3.3. Resistance-temperature coefficient for positive temperature coefficient thermistor

2.3. RESISTANCE TEMPERATURE SENSOR BRIDGE CIRCUITS

Thermistors are modulating transducers producing a resistance change. For laboratory work they are normally used in a Wheatstone bridge circuit, which must include some form of bridge balancing arrangement similar to that shown in figure 3.4. Bridge circuits may be dispensed within less demanding applications. While the resistance changes exhibited by a metallic resistance temperature device (RTD) are reasonably linear, those shown by semiconductor thermistors are markedly nonlinear. In both cases the resistance changes are large. Even if the sensor output is linear, the out-of-balance voltage measured using a bridge circuit is not necessarily linear for large changes in sensor resistance. Take, for example, the case of a 500 Ω platinum resistance thermometer, which exhibits a 100 Ω resistance change over its design temperature range. If the sensor is included in a bridge with four equal arms, the out-of-balance voltage will be very nonlinear as a function of temperature. However, if the fixed resistors R_1, R_2 (see figure 3.4) are of considerably higher resistance (about × 10 is normal) than R_3 and R_4 and if care is taken to balance the bridge at the middle of its design temperature range rather than at one end, reasonable linearity may be achieved.

Resistance thermometer bridges may be excited with either AC or DC voltages. The current through the sensor is usually in the range from 1 to 25 mA. This current causes I^2R heating to take place, which raises the temperature of the thermometer above that of its surroundings and causes a so-called self-heating error to occur. The magnitude of this error depends on the heat transfer conditions (for instance, the conductivity of the surface to which the sensor is attached, and the presence or otherwise of a fluid flow). However, it is not often significant in automotive engineering where an accuracy of ±1 °C is generally sufficient.

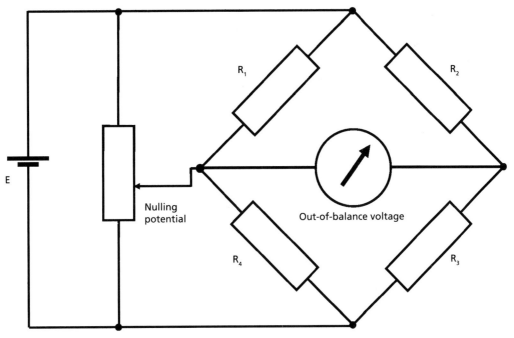

Figure 3.4. Bridge circuit with balancing (nulling) adjustment

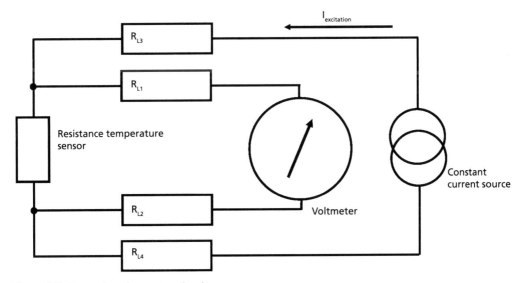

Figure 3.5. Four-wire ohmmeter circuit

An alternative to the classical bridge techniques for conditioning RTD sensors is the four-wire *ohmmeter* technique shown in figure 3.5. This is widely used with digital data acquisition systems, where any sensor nonlinearity is corrected in the computer software. A precision current source is used, so resistance changes in the two connecting wires *L3* and *L4* have no effect on the sensor current I_{ex}. The resistance of the four lead wires *L1*, *L2*, *L3*, and *L4* is shown as R_{L1}, R_{L2}, R_{L3}, and R_{L4} in figure 3.5. A high-impedance voltmeter is used, typically with a FET input of > 200 MΩ. This ensures that the currents through *L1* and *L2* are negligible, as are the lead-wire resistance errors.

3. THERMOCOUPLES

A thermocouple is a self-generating transducer comprising two or more junctions between dissimilar metals. The conventional arrangement is shown in figure 3.6a, and it will be noted that one junction (the "cold junction") has to be maintained at a known reference temperature, for instance, by surrounding it with melting ice. The other junction is attached to the object to be measured.

Thermocouple materials are broadly divided into two arbitrary groups based upon cost. The groups are known as the *base metal* and *precious metal* thermocouples. The most commonly used industrial thermocouples are specified by type letters as shown in table 3.1.

The arrangement of figure 3.6a is inconvenient because of the layout of leads and the need for a reference temperature. A more practical scheme is shown in figure 3.6b. The two wires are laid out

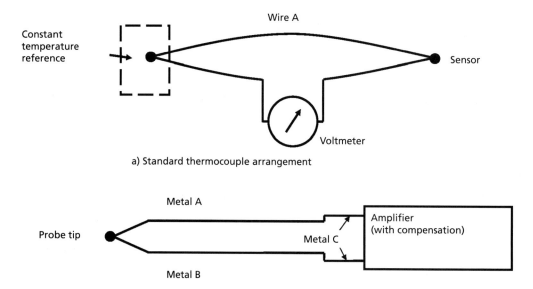

Figure 3.6. Thermocouple circuits

side by side and are connected to a voltage measuring circuit. The junctions between the two wires and the voltmeter do not cause any error signal to appear so long as they are at the same temperature. Since there is no proper reference junction with this approach, the system is liable to give an erroneous output if the temperature of the surrounding environment changes. This is avoided by the use of so-called *cold junction compensation* (see section 3.1) in which the characteristics of the signal conditioning amplifier are modified by including a thermistor (see section 3.2) in the circuit.

The arrangement shown in figure 3.6b is almost universally applied whenever thermocouples are used. The two wires are often enclosed within a tube or flexible sleeve of stainless steel or copper for protection, although this increases the time constant of the system.

The main advantages of thermocouples are their wide temperature range, nominally from −180 to +1200 °C for a Chromel/Alumel device, and their linearity. They are more expensive than thermistors and are therefore mainly used for prototype and experimental work rather than for production vehicles. Table 3.1 gives the characteristics of some of the most common commercially available devices.

If a short section of tubing is made of butt-welded thermocouple materials and inserted into a pipeline, the temperature of a fluid flowing inside the pipe may be measured nonintrusively. Ready-made sensors of this type are available commercially and are often used in automotive engineering for experimental work involving the measurement of fuel and coolant temperatures.

3.1. THERMOCOUPLE COMPENSATION

As noted earlier, it is not normally practical to have thermocouple cold junctions maintained at a controlled reference temperature. However, with the cold junctions at ambient temperature, which may change, some form of cold junction compensation is required. Consider the arrangement in figure 3.7, which shows a thermocouple with its measuring junction at temperature T °C and its cold junction at ambient temperature T_a. The thermocouple output is the voltage $V_{(T_a - T)}$ but what is required is the output which would be produced if the cold junction was at 0 °C, that is, $V_{(T_0 - T)}$. Thus a voltage $V_{(T_0 - T_a)}$ must be added to correct the output signal:

$$V_{(T_0 - T)} = V_{(T_a - T)} + V_{(T_0 - T_a)}. \qquad (3.5)$$

The voltage $V_{(T_0 - T_a)}$ is called the *cold junction compensation voltage*, and it is provided automatically by the circuit of figure 3.7, which includes a thermistor R_4 as part of the bridge. The other resistors R_1, R_2, and R_3 are temperature stable. The bridge is first balanced with all the components at 0 °C. As the ambient temperature is changed away from 0 °C, an unbalance voltage will appear across AB. This voltage is scaled by selecting R_4 such that the unbalance voltage across AB equals $V_{(T_0 - T_a)}$ in equation 3.5.

3.2. MULTIPLE THERMOCOUPLE ARRANGEMENTS

Several thermocouples may be connected in series or parallel as shown in figure 3.8 to achieve useful functions. The series arrangement (see figure 3.8a) is used mainly as a means of enhancing sensitivity. All the measuring junctions are held at one temperature, and all the reference junctions at another. This arrangement is often called a *thermopile,* and for *n* thermocouples gives an output *n* times as great as that which can be obtained from a single couple. A typical commercially available Chromel/Constantan thermopile has 25 junctions and produces about 0.5mV/°C.

Table 3.1. Thermocouples

Type	Conductors (positive conductor first)	Accuracy	Output for indicated temperature (cold junction at 0 °C)	Service temperature range
B	Platinum: 30% Rhodium alloy Platinum: 6% Rhodium alloy	0 to 1100 °C ±3 °C 1100 to 1550 °C ±4 °C	1.24 mV at 500 °C	0 to 1500 °C
E	Nickel: Chromium/Constantan	0 to 400 °C ±3 °C	6.32 mV at 100 °C	−200 to 850 °C
J	Iron/Constantan	0 to 300 °C ±3 °C 300 to 850 °C ±1%	5.27 mV at 100 °C	−200 to 850 °C
K	Nickel: Chromium/ Nickel: Aluminum (Chromel/Alumel)	0 to 400 °C ±3 °C 400 to 1100 °C ±1%	4.1 mV at 100 °C	−200 to 1100 °C
R	Platinum: 13% Rhodium/Platinum	0 to 1100 ±1 °C 1,100 to 1400 ±2 °C 1,400 to 1500 ±3 °C	4.47 mV at 500 °C	0 to 1500 °C
S	Platinum: 10% Rhodium/Platinum	As type R	4.23 mV at 500 °C	0 to 1500 °C
T	Copper/Constantan	0 to 100 °C ±1 °C 100 to 400 ±1%	4.28 mV at 100 °C	−250 to 400 °C

Notes:
Type B: Best life expectancy at high temperatures
Type E: Resistant to oxidizing atmospheres
Type J: Low cost, general purpose
Type K: General purpose, good in oxidizing atmospheres
Type R: High temperature, corrosion resistant
Type S: As R
Type T: High resistance to corrosion by water

The parallel arrangement shown in figure 3.8b generates the same temperature as a single couple if all the measuring and reference junctions are at common temperatures. If the measuring junctions are at different temperatures and the thermocouples all have the same resistance, the output voltage is the average of the individual voltages. The temperature corresponding to the output voltage is the mean temperature (as long as the thermocouples are linear over the measurement range).

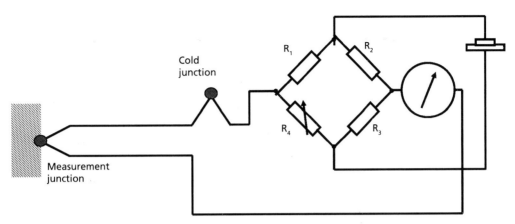

Figure 3.7. Thermocouple bridge circuit with cold junction compensation

4. BIMETALLIC TEMPERATURE SENSORS

Bimetal strips are made by bonding together two metals with different coefficients of thermal expansion. Typical materials used are brass and Invar. As the temperature of the bimetallic component changes the brass side expands or contracts more than the Invar, resulting in a change of curvature.

If two metal strips A and B with coefficients of thermal expansion α_A and α_B are bonded together as shown in figure 3.9, a temperature change will cause differential thermal expansion to occur. If it

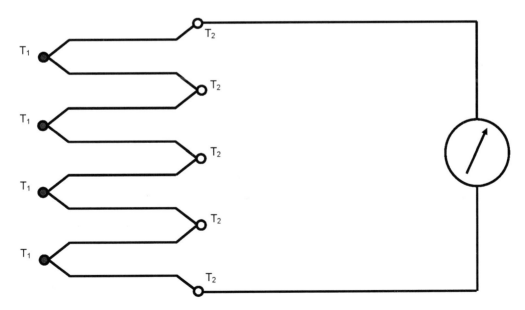

Figure 3.8a. Multiple junction thermocouples: thermopile

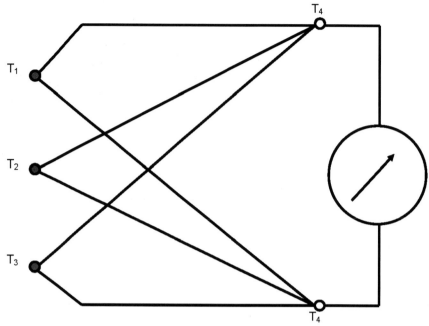

Figure 3.8b. Thermocouple averaging arrangement

is unrestrained the strip will deflect to form a uniform circular arc. Analysis [1] gives the radius of curvature ρ as

$$\rho = \frac{t[3(1+m)^2 + (1+mn).(m^2 + 1/(mn))]}{6(\alpha_A - \alpha_B).(T_2 - T_1).(1+m)^2}, \qquad (3.6)$$

where ρ is the radius of curvature,

t_A, t_B are the thicknesses of strips A and B,

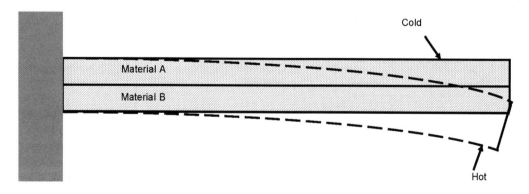

Figure 3.9. Bimetallic strip

t is the total strip thickness,
n is the ratio of elastic moduli, E_B/E_A
m is the thickness ratio, t_B/t_A, and
$T2 - T1$ is the temperature rise.

In most practical cases $t_B/t_A \approx 1$ and $(n + 1)/n \approx 2$, giving

$$\rho \approx \frac{2t}{3(\alpha_A - \alpha_B).(T_2 - T_1)}. \tag{3.7}$$

Equations 3.6 or 3.7 can be combined with the appropriate strength of materials expressions to calculate the deflections of most shapes of bimetal component, or of the forces developed by partially or completely restrained structures.

Bimetallic devices are used for low-cost temperature sensing and as "on-off" elements in temperature control systems. They are often used as overload cutout switches for electric motors. Their use in combination with a conductive plastic potentiometer has been reported for an automotive temperature sensing application [2].

5. PN JUNCTION SENSORS

PN junctions in silicon have become popular as temperature sensors due to their very low cost. Figure 3.10 shows the forward bias characteristic of a silicon diode. It is well known that a voltage V_f has to be applied across the junction before a current will flow. For silicon V_f (which is often termed the *diode voltage drop*) is of the order of 600–700 mV. V_f is temperature dependent, and is very nearly linear over the temperature range from –50 to +150 °C. This covers the range of temperatures usually required in automotive engineering. The voltage V_f has a temperature characteristic that is essentially the same for all silicon devices of about –2mV/°C.

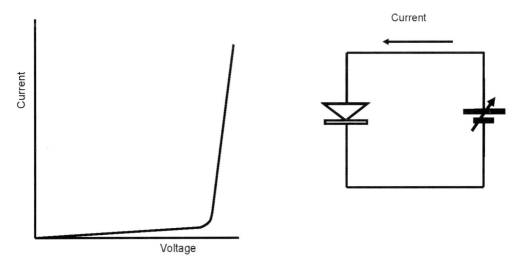

Figure 3.10. Silicon diode forward bias characteristic

A typical circuit is shown in figure 3.11. Transistors are often used instead of diodes, with the base and collector terminals connected together. For the best performance a constant current should flow through the diodes, but in practice the error incurred by driving the circuit from a constant voltage is small.

There is one major disadvantage to using diodes as temperature sensors in control applications, which is that they are not fail-safe. If a diode temperature sensor is used to control a heater, any breakage of the diode wires will be interpreted by the controller as low temperature. More power will then be applied to the heater, resulting in an uncontrolled runaway.

6. LIQUID CRYSTAL TEMPERATURE SENSORS

A number of liquids (mainly organic) can be made to exhibit an orderly structure, in which most or all of the molecules are aligned in a common direction. The structure can be altered by electric or magnetic fields. Most people are familiar with the liquid crystal displays used in watches and calculators, which use compounds sensitive to electric fields. Less well known, however, is the fact that some liquid crystal materials are temperature sensitive. One application, which may be familiar, is the use of cholesteric compounds as medical thermometers in the form of a flexible plastic strip that is pressed against the skin to measure its surface temperature. The compounds involved have molecular structures similar to cholesterol, and for this reason are called "cholesteric." Cholesteric liquids form a

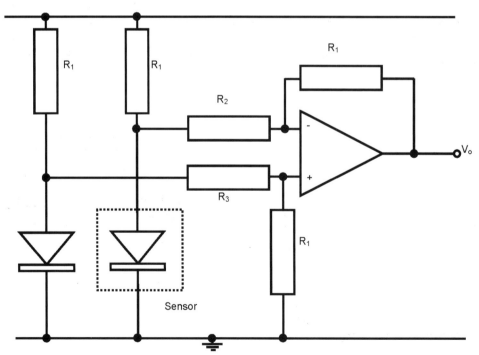

Figure 3.11. Signal processing circuit for PN junction temperature sensor

helical structure, and this gives them their optical properties. The plane of polarization of light passing through the compound is rotated by as much as 1000° per millimeter of path length.

The structure can be enhanced by confining the cholesteric liquid between two parallel sheets of a suitable plastic. The choice of polymer for the plastic is based on two requirements. These are, first, that it must be optically transparent, and, second, that it is sufficiently chemically active to bond to the liquid crystal molecules adjacent to the polymer and maintain their axes in the correct orientation.

When used for temperature measurement, the liquid crystal is confined between two polymer sheets as described above, a few tens of microns apart. The surface of one plastic sheet is given a reflective coating as shown in figure 3.12. In figure 3.12a, a light ray enters the sandwich and is reflected back. Since the liquid crystal is in its ordered state, the reflected ray interferes destructively with the incident light and the sensor appears opaque.

In figure 3.12b the liquid crystal has been raised to a temperature at which its ordered structure has broken down due to thermal agitation. The temperature at which this occurs depends on the exact molecular structure. It is reasonably easy to create compounds tailored to specific temperature changes. If a ray of light enters the sandwich in this condition, it is reflected back in the usual way and the sensor appears to be transparent (the color of the backing material is usually visible).

If polarized light is used, this technique can detect temperature changes as small as 10^{-3} °C. If ordinary white light is used, the resolution is of the order of 0.1 °C. The advantages of liquid crystal temperature sensors are that they are relatively cheap and immune to electromagnetic interference. This last characteristic may make them suitable for automotive applications for which radio-frequency interference (RFI) problems are too severe for a conventional electronic sensor to be used. It is possible to interrogate a liquid crystal temperature sensor remotely, by means of a fiber-optic link.

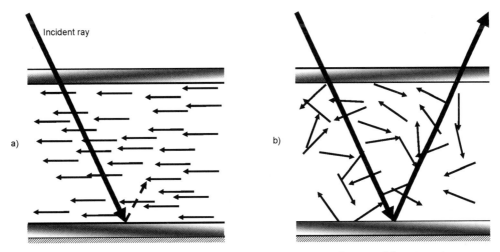

Figure 3.12. Liquid crystal temperature sensor: (a) destructive interference, (b) no interference

7. INFRARED EMISSION AND PYROMETRY

The amount and the wavelength of radiation emitted by a body are functions of its temperature. This dependence on temperature of the characteristics of radiation is used as the basis of a noncontact temperature measurement technique in which the sensors used are known as *radiation thermometers*. Automotive applications of these devices have been reported, for example, in tire condition monitoring; see reference 2.

The total power of radiant flux of all the wavelengths $E_o(T)$ emitted by a black body of area A is proportional to the fourth power of the temperature of the body in Kelvin:

$$E_o(T) = \sigma A T^4, \qquad (3.8)$$

where σ is the Stefan-Boltzmann constant, which has the value 5.67032×10^{-8} W/m²K⁴. Most radiation thermometers are based on this law, since if a sensing element of area A at a temperature T_1 receives radiation from an object at temperature T_2, it will receive heat at a rate $A(T_2)^4$ and will emit heat at a rate $\sigma A(T_1)^4$. The net rate of heat gain is therefore $(T_2^4 - T_1^4)$.

If the temperature of the sensor is small compared to that of the source, T_1^4 may be neglected in comparison with T_2^4.

The above discussion applies to perfectly black bodies with an emissivity ε of unity. "Real" objects have nonunity emissivities, and a correction must be made for this. The total radiant flux emitted by an object of area A and emissivity ε is

$$E(T) = \sigma \varepsilon A T^4. \qquad (3.9)$$

However, $E(T)$ is numerically equal to that emitted by a *perfect* black body at a different temperature T_{app}, the *apparent temperature* of the body, that is, the energy flux $E_0(T_{app})$ where

$$E_0(T_{app}) = \sigma A T_{app}^4. \qquad (3.10)$$

Equating (3.9) and (3.10) gives

$$T = \frac{T_{app}}{\sqrt[4]{\varepsilon}}. \qquad (3.11)$$

Radiation thermometers generally consist of a cylindrical body made from aluminum alloy or plastic. One end of the body carries a lens, which focuses energy from the target onto a detector within the tube. The lens may be made from glass, germanium, zinc sulfide, sapphire, or quartz, depending on the wavelength. The heat detectors within the instrument are generally thermocouples, thermistors, or PN junctions.

8. SOLID AND LIQUID EXPANSION TEMPERATURE SENSORS FOR AUTOMOTIVE USE

All water-cooled engines control the coolant circulation to avoid overcooling. This is often done quite crudely, using a temperature-sensitive valve called a *thermostat*. A thermostat is not just a simple sensor that provides information to be used elsewhere, but an integrated, self-contained set-point control system that comprises a sensing element, a control system, and an actuator.

It seems likely that in the future automotive engine cooling will be placed under more sophisticated control, perhaps by means of a variable-speed electric drive to the water pump and one or more electronic temperature sensors placed around the engine [3].

Although thermostats are not strictly temperature sensors they are worth including here, since they can be considered to be temperature transducers (transducing temperature to displacement). There are two kinds of thermostats in common use: the *bellows* and the *wax element* types.

8.1. BELLOWS THERMOSTATS

The operating element of bellows thermostats is a sealed flexible metal bellows, partly filled with a liquid that has a lower boiling point than that of the engine coolant fluid. Alcohol, ether, or acetone are commonly used. Air is excluded from the bellows, which contains only the liquid and its vapor. The pressure inside the bellows is therefore the vapor pressure of the liquid. This varies with temperature, being equal to atmospheric pressure at the boiling point of the liquid. The pressure is less below boiling point, and more at higher temperatures. The extension of the bellows is thus temperature dependent, and this is used to control the coolant flow. A poppet-type valve is attached by a stem to the top of the bellows, as shown in figure 3.13a, and varies a circular opening in the flange to control the flow of water.

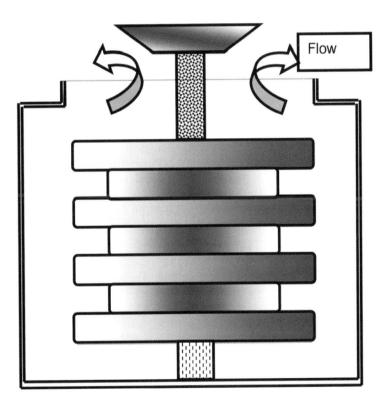

Figure 3.13a. Bellows thermostat

Complete sealing is not necessary and indeed is undesirable, since it would lead to air being trapped beneath the valve when the system is filled. A small hole is normally drilled in the valve to act as a vent. A loose-fitting *jiggle-pin* prevents this hole from becoming clogged.

The free length of the bellows is such that when internal and external pressures are equal, the valve will remain open. Thus if the bellows develops a leak, the valve will remain open; that is, the thermostat is fail-safe.

8.2. WAX ELEMENT THERMOSTATS

A wax element thermostat depends for its operation upon the considerable change in volume that occurs when certain types of wax melt. The operating element is a metal cylinder filled with such a wax, into which is inserted a thrust pin, as shown in figure 3.13b. A flexible rubber sleeve surrounds the pin and is sealed to the top of the capsule to prevent the wax escaping. A poppet valve is again used and is held in the shut position by a spring when cold. When the thermostat is heated the wax melts and expands, pushing the thrust pin out of the capsule and opening the valve.

The useful life of this type of thermostat is claimed to be in excess of 100,000 km. The opening temperature tends to increase, however, due to deterioration of the rubber sleeve.

Failure can take place in two ways. The most common mode is for a leak to develop, allowing the wax to escape and the valve to remain closed. The vehicle then overheats, so this device is not as fail-safe as the bellows type.

Failure can also occur due to a leak developing in the rubber sleeve below the thrust pin. In this case the valve will stick open.

A vent hole and jiggle-pin are also provided for the reasons given in section 8.1.

9. HEAT FLUX GAUGES

It is sometimes necessary to make local measurements of convective, radiative, or total heat transfer rates. In automotive engineering this is most likely to occur in engine and cooling system research. This requirement has led to the development of a class of sensors known as *heat flux gauges*.

One common type of heat flux gauge has the general form shown in figure 3.14. Two temperature measuring elements (usually thermocouples) are physically separated by a thermal insulator with

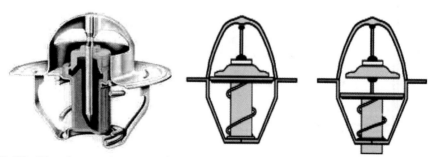

Figure 3.13b. Wax element thermostat (*Source:* Standard-Thomson Corporation, 152 Grove Street, Waltham, MA)

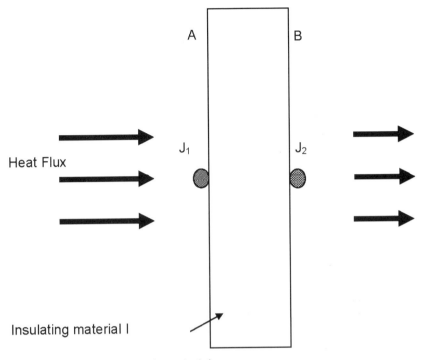

Figure 3.14a. Heat flux gauge operating principle

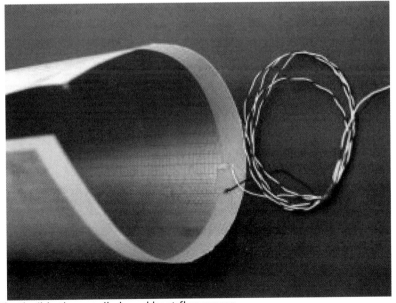

Figure 3.14b. Flexible thermopile-based heat flux sensor

known characteristics. When heat energy begins to pass through surface A, the thermocouple J1 generates a small voltage. Since the heat has to pass through the thickness of insulating material I to reach the second thermocouple on surface B, a different voltage is generated by J4. The differential voltage developed across J1 and J2 is proportional to their temperature difference. If the characteristics of the insulator I are known, the heat transfer rate may be obtained as a function of voltage.

Heat flux gauges of this type are fabricated as a string of thermocouples on a flexible backing. Up to 30 thermocouple junctions are placed on each side of the insulator to increase the sensitivity of the device as discussed in section 3.2. Heat flux gauges of the type described in figure 3.14 somewhat resemble strain gauges and are bonded to the surface to be measured in a similar fashion. However, strain gauges are usually considered to be disposable since they are relatively cheap. Heat flux gauges currently cost from $300 to $500 each, although the price is falling, and it is normal to try and remove them for reuse.

10. SUMMARY AND CONCLUSIONS

The types of temperature sensor required in the automotive industry may be divided into two categories: those used for research and development, and those fitted to production vehicles. The former encompasses almost all of the temperature measurement methods available, and a complete treatment is beyond the scope of this book, although descriptions of the most commonly used devices and techniques have been given. The second class is of more interest to the automotive engineer since temperature measurement is essential for a number of control functions, notably that of electronic engine management where inlet air temperature is required.

Up to about a decade ago, most of the temperature sensors used on road vehicles were of the bimetallic variety. This has gradually changed, and nowadays thermistors are much more common. These are usually of the bead type, although it seems likely that thick-film thermistors will become standard because of their low cost and the fact that they can be integrated with other circuitry in the form of a thick-film hybrid.

A number of emerging applications such as tire condition and prop shaft joint monitoring may require noninvasive temperature measurement techniques, and infrared emission sensors may be the most suitable for this purpose.

Silicon diode junctions exhibit useful thermal characteristics, as discussed in section 5. These are likely to become common in applications where temperature measurement is required as part of an integrated silicon device. This might be the case in a micromachined silicon pressure sensor where pressure and temperature measurements are frequently required together. If the expense of designing a custom silicon device is considered to be worthwhile, temperature sensing in the form of a PN junction can be included at very little extra cost.

As the amount of electronics on a vehicle increases, the requirement for temperature sensing is also likely to increase. This may be because temperature data is required to control a mechanical function. It seems more likely, however, that temperature measurements will be increasingly needed to provide thermal compensation of the electronic systems themselves.

REFERENCES

1. Doebelin, E. O. 1990. *Measurement systems application and design*. 4th ed. New York: McGraw-Hill.
2. Hutchinson, M. 1989. An infra-red tyre condition monitor. *Proceedings of the 7th International Conference on Automotive Electronics*. London: I. Mech. E. paper C391/009.
3. McBride, J. W., and M. J. Reed. 1989. An electrically-driven automotive coolant pump. *Proceedings of the 7th International Conference on Automotive Electronics*. London: I. Mech. E. paper C391/054.

CHAPTER 4

AUTOMOTIVE AIRFLOW SENSORS

John Turner

1. INTRODUCTION

The subject of flow measurement can be subdivided into vector flow, where the direction and the magnitude of flow are sensed, volume flow, and mass flow rate. In automotive engineering mass flow rate is the most important. Information on the rate at which air mass flows into an internal combustion engine is essential if the air-fuel ratio and therefore the combustion process is to be accurately controlled.

In older carburetor-type engine designs, airflow across the end of the carburetor jet was sufficient to draw fuel into the air stream and hence to supply the engine. By using jets of different diameter it is possible to adjust air-fuel ratio so that it is approximately correct across the hot working range of the engine. Additional or variable jets are used for acceleration or cold-start conditions.

For at least the last two decades, however, the use of carburetors in automotive applications has almost ceased, since carburetors alone are incapable of providing fine enough control of the air-fuel ratio to meet the critical demands of exhaust catalysts. The almost universal design approach now is to use fuel injection together with a separate means of measuring the mass of air entering the engine. Thus, sensors for measuring airflow have become critically important to the automotive engineer.

Flow sensors are also important in many other engineering applications, particularly those where the measurements are used to control a manufacturing process. The first method of airflow sensing to be used on a vehicle was an indirect approach where inlet manifold pressure was measured using a pressure transducer (see chapter 3) and the data used to estimate air mass flow rate from a knowledge of the volumetric flow rate of the engine, its rotational speed, and the air density (estimated from air temperature [1]). This approach is known as the speed-density method [2]. The measurement accuracy obtained is adversely affected by pressure variations caused by resonances in the manifold, by the effect of flow reversal in some types of engine, by manufacturing variability, and by engine wear.

For the reasons discussed above, a direct method of measuring air mass flow rate into an engine was sought.

2. MOVING-VANE AIRFLOW (VAF) SENSORS

The first (and still widely used) transducer is the vane airflow meter developed by Bosch. A diagram of this device is shown in figure 4.1. A vane airflow sensor is located upstream from the throttle and measures the mass flow rate of air entering the engine by the deflection of a spring-loaded mechanical flap. The flap is pushed open by an amount proportional to the mass flow rate of air entering the engine. Deflection of the flap is sensed by a rotary potentiometer. The greater the airflow, the wider the flap opens, and the larger the resulting change in potentiometer resistance.

The vane airflow sensor also contains a safety switch for the electric fuel pump relay. Airflow into the engine activates the pump. A sealed idle-mixture screw is also located on the airflow sensor. This controls the amount of air that bypasses the flap, and consequently the air-fuel mixture ratio.

The major problem with vane airflow sensors is their intolerance of air leaks. A vacuum leak downstream of the VAF sensor allows "unmetered" air to enter the engine. This additional air can cause drivability problems. The exhaust-gas oxygen (EGO) sensor can compensate for small air leaks once the engine warms up and goes into closed-loop control mode, but not for large air leaks. The oxygen sensor is of course downstream of the combustion process, which means that air leaks can cause hesitation when the throttle is suddenly opened.

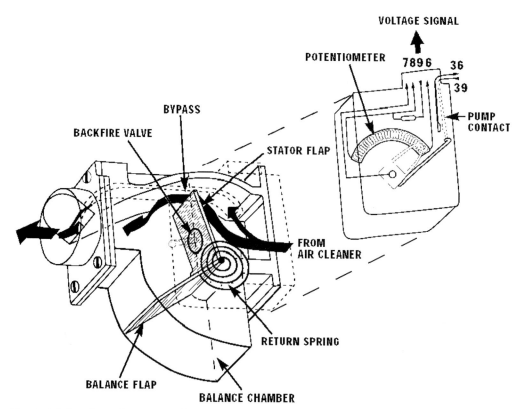

Figure 4.1. Bosch moving-vane airflow sensor

Vane airflow sensors are also vulnerable to dirt. Unfiltered air passing through a torn or badly fitting air filter can allow dirt to build up on the flap shaft, causing the flap to bind or stick.

Backfiring in the intake manifold can force the flap backwards violently, often bending or breaking the flap. Some sensors have a "backfire" valve built into the flap to protect the flap in case of a backfire by venting the explosion. However, the antibackfire valve can itself cause problems if it leaks, as it can affect the air-fuel mixture.

3. HOT-WIRE AND HOT-FILM FLOW TRANSDUCERS

Mass airflow (MAF) sensors are used on many multiport fuel-injected engines to measure the volume of air entering the engine. This is necessary to enable the engine management system to calculate and maintain the proper air-fuel ratio for optimum performance and emissions. The MAF sensor is located in the air duct between the air cleaner and throttle body. In this location it can measure all the air that is being drawn into the engine, and reacts almost instantly to changes in throttle position and engine load.

A hot-wire anemometer consists of an electrically heated wire or metallized-film probe, which is placed in the moving fluid as shown in figure 4.2. The rate at which thermal energy is lost from the wire depends on the flow rate as shown in the following equation:

$$I^2 R = K_h A (T_w - T_f)$$ (4.1)

where I is current

R is the electrical resistance of the wire (which is usually platinum)

K_h is the wire or film coefficient of heat transfer*

A is the probe's heat transfer area

T_w is the wire (or film) temperature

---

* K_h is a constant for a probe of given material and geometry.

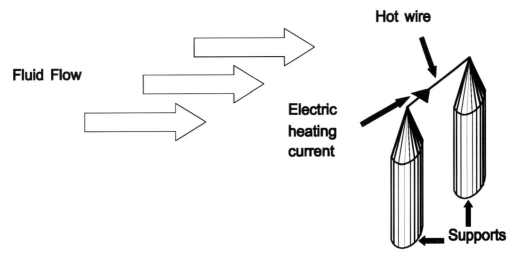

Figure 4.2. Hot wire/film transducer

T_f is the temperature of the moving fluid

The film coefficient of heat transfer K_h is a function of the fluid velocity V_f and two constants, K_{h0} and K_{h1}, as shown in equation 4.2:

$$K_h = K_{h0} + K_{h1}\sqrt{v_f} \qquad (4.2)$$

The output of a hot-wire anemometer is also a function of the direction of flow. A simple transducer such as the device shown in figure 4.2 has a maximum output when the flow is orthogonal to the wire, and reduces in an approximately cosine fashion as the probe is rotated, reaching a minimum of around 20% of the maximum output when the wire is longitudinally aligned with the flow. The hot wire (or hot film) has a known resistance/temperature characteristic, and is made to form one arm of a bridge circuit. Either the heating current is maintained at a constant value or a feedback system is used to keep the temperature of the wire constant. This second approach has the advantage that the heating current is then proportional to the flow.

There are two basic varieties of automotive MAF sensors: hot wire and hot film.

Figure 4.3 shows a typical commercial automotive device. As air flows past the sensing element, it has a cooling effect. As shown earlier, this increases the current needed to keep the sensing element at a constant temperature. The cooling effect varies directly with the temperature, density, and humidity of the incoming air. Thus, electrical current is proportional to the air "mass" entering the engine.

One problem with automotive hot-wire and hot-film devices is that their accuracy can become degraded due to dirt, soot, and so forth on the wire. A self-cleaning cycle is often provided where the platinum wire is heated to 1000 °C for one second after the engine is shut down to burn off contaminants that might otherwise interfere with the sensor's ability to measure incoming air mass accurately.

4. VORTEX-SHEDDING FLOWMETERS

Moving-vane and hot-wire or hot-film mass airflow sensors are the most common means of measuring the amount of air entering an engine. However, vortex-shedding devices have also been used on production vehicles: The first such use was described in 1976 [3,4].

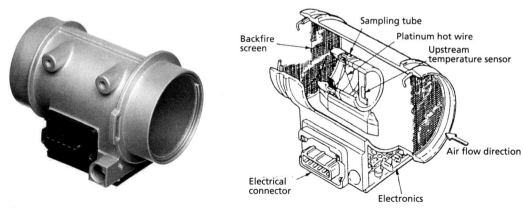

Figure 4.3. Hot-wire or hot-film mass airflow (MAF) sensor

The phenomenon of vortex shedding downstream from a bluff body immersed in a steady flow is well known. It can arise unintentionally, as was the case at the Tacoma Narrows suspension bridge in the United States, which collapsed due to wind-induced oscillation in 1940 [5]. More usefully, vortex shedding is the basis of the vortex-shedding flowmeter, shown in the cross section in figure 4.4. When the Reynolds number[†] exceeds around 10000 vortex shedding occurs reliably, and the shedding frequency f is given by equation 4.3:

$$f = \frac{N_{st} V_f}{d}, \qquad (4.3)$$

where V_f is the fluid velocity, d the characteristic dimension of the shedding body, and N_{st} the Strouhal number, which is experimentally determined. By careful design N_{st} can be kept nearly constant over a large range of Reynold's number N_{Re} and thus flow rate, making f proportional to V. Vortex-shedding flowmeters are therefore pseudo-digital devices, in which the flow is measured by counting the vortex-shedding rate (see figure 4.4b).

Various methods are employed to count the vortex-shedding rate. The vortices cause local pressure changes (and hence force changes) on the bluff body, and piezoelectric or strain-gauge transducers can be used to detect these. Hot-film transducers (see previous section) buried within the vortex shedder can detect the local flow fluctuations caused by the vortices. Another common approach is to use the effect of the vortices on beams of ultrasound passed through the fluid. A mechanical approach is sometimes used, in which vortex-induced differential pressures cause oscillation of a small caged ball, the motion of which is detected by a magnetic proximity pickup. A further method is to direct the vortices into a chamber from which they can apply force to a metal foil mirror, causing the mirror to oscillate at the vortex-shedding frequency. Optical sensing methods can then be used as shown in figure 4.5.

5. ULTRASONIC FLOWMETERS

In an ultrasonic flowmeter, small-magnitude pressure disturbances at a high frequency (normally 20 to 50 kHz) are propagated through the fluid at the speed of sound. Figure 4.6 shows a typical arrangement with transducers separated by a distance L. If the fluid is flowing with velocity V_f and the speed of sound in the stationary fluid is c, the propagation speed v will be the sum of the two, as shown in equation 4.4:

$$v = v_f + c. \qquad (4.4)$$

Figure 4.6a shows the most direct implementation of this approach. When the flow velocity is zero, the transit time t_o is given by

$$t_o = \frac{L}{c}, \qquad (4.5)$$

where L is the distance between transmitter and receiver as shown in figure 4.6. In water the speed of sound $c \approx 1500$ m/s, so if $L = 10$ cm, $t_o \approx 70$ μs.

[†] The Reynolds number N_{Re} for a flow through a circular pipe is defined as

$$N_{Re} = \frac{\rho u d}{\eta} = \frac{u}{v} d,$$

where ρ is the density of the fluid, u is mean fluid velocity, d is the pipe diameter, η is the (absolute) dynamic fluid viscosity, and v is the kinematic viscosity of the fluid (ν = η/ρ). Laminar flow is experimentally found to occur for Re < 30. At larger Reynolds numbers, flow becomes turbulent.

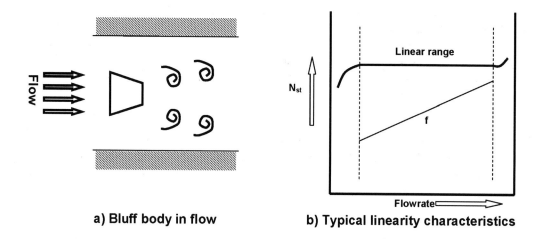

a) Bluff body in flow   b) Typical linearity characteristics

Figure 4.4. Vortex-shedding flow transducer

If the fluid is moving at a velocity v_f, the transit time t becomes

$$t = \frac{L}{c + v_f} = L\left(\frac{1}{c} - \frac{v_f}{c^2} + \frac{v_f^2}{c^3}\right) \approx \frac{L}{c}\left(1 - \frac{v_f}{c}\right). \tag{4.6}$$

If the change in transit time $\Delta t = t_o - t$, then

$$\Delta t \approx \frac{L v_f}{c^2}. \tag{4.7}$$

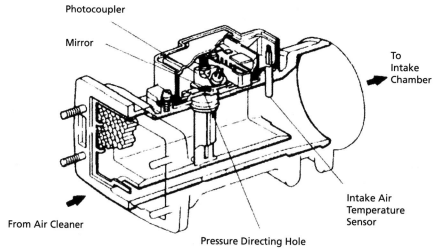

Figure 4.5. Automotive vortex-shedding mass airflow transducer

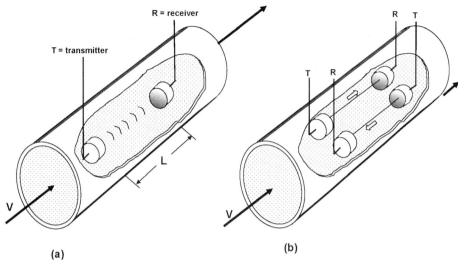

Figure 4.6. Ultrasonic flowmeter

Thus if L and c are known, measurement of Δt allows v_f to be calculated.

While L can usually be considered a constant, c varies with temperature for most fluids. Since c appears as c^2 in equation 4.7, the error caused by a temperature change may be significant. Also, Δt is quite small, since v_f is normally a small fraction of c. For example, if v_f is 10 m/s, L = 10 cm and c = 1500 m/s, then $\Delta t \approx 0.4$ μs, a short increment of time which can be hard to measure.

The arrangement shown in figure 4.6a does not directly measure t_o, and for this reason the modified arrangement shown in figure 4.6b is often preferred. If t_1 is the transit time with the flow, and t_2 is the transit time against the flow, then equation 4.8 applies:

$$\Delta t = t_2 - t_1 = \frac{2VL}{c^2 - V^2} \approx \frac{2VL}{c^2} . \tag{4.8}$$

This Δt is twice as large as before and so is easier to measure. The quantity Δt is also a quantity that may be directly measured (i.e., there is no need to know a "starting time," t_o).

Claims have been made for measurement accuracies of air velocity of better than 1% over a typical inlet mass airflow range of 0.01 to 0.2 kg/sec [6]. Below this bottom limit the onset of laminar flow introduces major inaccuracies. Beyond the upper limit measurement becomes difficult due to the onset of high turbulence.

Ultrasonic transducers can be operated across the temperature range from –40 to +140 °C (this is a typical thermal range for engine sensors), provided the air temperature does not undergo sudden changes. The measuring tube and its associated electronics require careful design for satisfactory operation. The ultrasound transducers have a very high impedance, and so the potential for electromagnetic interference is high.

REFERENCES

1. Turner, J. D., and A. J. Pretlove. 1991. *Acoustics for engineers*. Basingstoke: Macmillan.
2. Westbrook, M. H., and J. D. Turner. 1994. *Automotive sensors*. Bristol: IOPP.
3. Cartnell, B. C., and F. L. Zeisler. 1988. An engine mass airflow meter. SAE paper 760017.
4. Joy, R. D. 1976. Airflow for engine control. SAE paper 760018.
5. Gordon, J. E. 1978. *Structures*. London: Penguin.
6. Rinolfi, R. 1977. Engine control system with ultrasonic mass air flowmeter. SAE paper 770855.

CHAPTER 5

COMBUSTION SENSORS

Peter Eastwood

1. INTRODUCTION

After more than a century of development, internal combustion engine technology may be considered mature. One view is that conventional powertrains are probably approaching the limits of refinement and may not be able to satisfy increasingly strict environmental legislation. Powertrains using fuel cells or other technology may be preferred in the future.

There is an alternative vision, which predicts many years of further development for the internal combustion (IC) engine. Modern signal processing techniques, microprocessors, neural networks, fuzzy logic, adaptive control, and smart sensors are all being used to enhance environmental performance. These developments are likely to be of much lower cost than the huge investment required by widespread adoption of fuel cells.

The engine in a modern motor vehicle is monitored by a large number of sensors and these sensors are interrogated, in turn, by a complex and sophisticated engine management system. This is known in the automotive industry as the engine control unit (ECU) or engine control module (ECM). The function of the ECU is to make decisions, principally about how best to burn the fuel—that is, how best to control the combustion. ECUs and their associated sensors exist because without them present-day customer expectations and legislative constraints would be unattainable. The three main areas defining powertrain performance are emissions, economy, and drivability. Of these, legislative or statutory constraints on tailpipe emissions are perhaps the most difficult to satisfy and thus attract the lion's share of resources.

Compromise is inevitable in the powertrain design process and is resolved by design optimization. This optimization must be maintained in the face of several perturbing factors:

- Variations in production
- Variations throughout vehicle life (over, say, 10 years or 100000 miles)

- Variations in vehicle environment (temperature, humidity, altitude, fuel composition)
- Variations in customer driving patterns or duty cycles

The ability of a system to continue to operate optimally in the face of external perturbing factors is what engineers mean by robustness. Obviously the engineer cannot test for every contingency. One benefit of closed-loop control is that an ECU should be able to maintain optimized performance indefinitely and automatically.

However, open-loop control dominates in many areas of engine management. This statement will be illustrated by examining a common example. On a diesel engine the ECU would ideally know when combustion starts (start of combustion, SOC). However, the ECU does not measure SOC. What is measured is the start of (fuel) injection (SOI). There is a time delay (the ignition delay) between SOI and SOC. This need not be an issue so long as the delay is well understood and closely controlled. For example, a consistent diesel fuel formulation (i.e., Cetane number) helps to ensure a consistent ignition delay. SOI thus replaces SOC as the controlled variable insofar as it acts as a kind of proxy or surrogate.

Ignition delay is also a function of engine operating point. For every combination of engine speed and torque, the appropriate SOI must be known so that the appropriate SOC can be estimated. These measurements are done during the engine development stage and invariably involve lengthy investigations. This laborious practice is known as engine mapping or, more confusingly, as calibration. The SOI for each operating point, once determined, is programmed into an erasable programmable read-only memory (EPROM) to form a three-dimensional lookup table or map. One axis of the map lists engine speed; the other has some parameter representing engine load (torque, fueling). This map resides in a computer memory so that, knowing the engine's load and speed, the ECU can find, with interpolation, the SOI for any particular operating point. The corresponding SOC should then be realized as planned.

One difficulty that emerges from our example is that ignition delay is not solely a function of the Cetane number or engine operating point. The temperature of the incoming air is also influential. If the air is cold, the fuel takes longer to vaporize and ignite, and then the ignition delay lengthens. The solution is to advance or retard SOI according to ambient temperature. If this is done correctly, SOC will then continue to occur at the appropriate moment, irrespective of ambient temperature–related variations in ignition delay. A vehicle is designed to operate in a range of ambient temperatures, typically –30 °C to +40 °C. A series of maps for SOI are, therefore, required in order to cope with varying seasons and climates, say, –30 °C, –15 °C, 0 °C, +20 °C, and +40 °C, with interpolation between these.

Ignition delay is also a function of barometric pressure, and a motor vehicle must function optimally at altitudes from sea level to the highest mountain road. To achieve this further maps are required.

The approach outlined above lacks elegance. First, and most obviously, there are the uncertainties inherent in any control system lacking feedback. For example, the inferred load, listed along one side of the map, was determined in the development stage. The following problems may be encountered:

- An engine may not develop the intended torque in customer use.
- Aging may change the characteristics of critical components such as the fuel injectors.
- The optimized values stored in each map are interrelated, something that can lead to "rippling" or "unraveling" effects [1]. A modification to SOI can, for example, affect the combustion and alter the rate of exhaust gas recirculation.

- The optimization process, being essentially empirical, requires extensive testing and validation; and when something is changed—such as a value in a map or the fuel injectors—development must begin again.
- Large numbers of maps are generally required. (A typical ECU may contain hundreds of maps, and no engineer can possibly become familiar with all of them.)

Extensive research is, therefore, underway to simplify and reduce the number of engine maps. Several fields of work related to the present review are worth mentioning. Virtual sensors may be used [2] wherein information is derived from other available data, perhaps involving the use of fuzzy logic. Algorithms have been developed to extract information not directly measured. Alternatively, model-based control may be used in which mathematical models are built of the powertrain [3].

Returning to the topic of reducing the number and complexity of maps, it will be apparent that combustion sensors offer considerable map-culling potential. Most current ECUs rely on sensors located outside the engine, monitoring the air that enters and the exhaust gas that leaves. Inferences about the combustion process itself are necessarily indirect. In addition most such sensors operate on long timescales compared to the combustion cycle. The possibilities for real-time control using existing transducers are thus restricted.

With this in mind, this chapter focuses on combustion (i.e., in-cylinder) sensors suitable for production vehicle use. The function of such sensors is to monitor the combustion but as an unfolding process. Some form of signal is made available, generally as a function of crank angle (CA). A high degree of temporal resolution is usually required—not less than one measurement per crank angle degree (CAD), and often more than this during the combustion event.

Before proceeding to the review proper, it is worth dwelling for a moment on two types of sensor that will not be described any further: chemical sensors and temperature sensors.

1.1. CHEMICAL SENSORS

Chemical sensors, of which the Lambda or oxygen sensor is certainly the best known, are not used for in-cylinder applications but are mounted in the exhaust stream. The response of a Lambda sensor is too slow for use as a combustion sensor. However, when mounted close to the exhaust ports, it is sufficiently fast to provide cycle-resolved information, such as detecting a difference between one cycle and the next. The heated exhaust gas oxygen (HEGO) sensor, formerly the exhaust gas oxygen (EGO) sensor, was developed and first used in closed-loop control more than a generation ago [4]. The universal exhaust gas oxygen (UEGO) sensor [5] is only now moving into serial production.

The oxides of nitrogen (NO_x) sensor [6] is a comparatively recent innovation in the motor industry. Exhaust sensors are an actively researched field and new devices continue to emerge, for example, silicon carbide Schottky diodes [7].

1.2. TEMPERATURE SENSORS

Temperature sensors might be thought to have an unacceptably slow response compared with the extreme rapidity of thermal transients within the combustion chamber. This is not so. Traditional beaded thermocouples are undoubtedly sluggish but thin-film thermocouples respond on timescales of a few microseconds. Mounted on spark plugs, or flush-mounted in cylinder heads, thin-film thermocouples have been shown to capture the combustion event and measure temperature changes throughout the cycle of a few tens of degrees centigrade [8]. Similar results were reported with platinum-resistance sensors [9]. These temperatures, of course, relate to the wall rather than the combustion zone. However, temperature sensors are unlikely to survive for long in these hostile locations. Combustion temperature is, in any case, available indirectly through optical measurements of the radiant emission or luminosity, as described later.

There are only three serious candidates for real-time in-cylinder control: optical sensors (luminosity); electrical sensors (based on ion current); and mechanical sensors (measuring pressure). For each case we shall consider three issues: (a) practical aspects of sensor design and construction; (b) operating principles and characteristics; and (c) actual applications. We shall cover the first two issues reasonably comprehensively. The literature contains numerous descriptions of sensor applications and thus only "tasters" are provided. A final table, however, directs the reader to more detailed investigations.

2. POTENTIAL APPLICATIONS FOR COMBUSTION SENSORS

The previous section has shown that engine management systems could be improved by in-cylinder sensors making direct measurements of the combustion process. Real-time control of the combustion process involves making an adjustment in one cycle and observing the effects of that adjustment on the following cycle. With this approach robustness could be greatly improved and optimum performance maintained. The uncertainties inherent in open-loop control can be avoided with less reliance on inferred information. The amount of engine mapping required would also be reduced.

2.1. GENERAL COMBUSTION SENSORS

Several parameters are used to form the traditional data set for combustion studies and are listed below. These measurements are mainly used to optimize engine performance in the development stage, although potential commercial applications do exist.

Start of combustion (SOC). Although SOI is known from the command signal sent to the fuel injectors, the ECU has no knowledge of the first stages of the burn. If the burn commences too early (for example, before the piston reaches top dead center or TDC), energy is wasted in opposing the piston motion. On diesel engines particularly, it would be useful for an ECU to know the true ignition delay.

Cylinder pressure. This is the key measure of engine performance, since it immediately reveals the force on the piston and thus the torque transmitted to the crankshaft. In forced-induction (i.e., turbo- or super-charged) engines cylinder pressure is a significant constraint on full-load performance. If the pressure is too high, cylinder-head damage can result. Cylinder pressure is available in principle through an examination of fluctuations in crankshaft speed, a method that competes with direct in-cylinder measurements [10]. The chief difficulties with this method are discarding torsional resonances, extracting information from a signal containing data from several cylinders [11], and rejecting electrical noise.

Mass fraction burned (MFB). This is the fraction of fuel burned normalized to the total quantity of fuel injected, expressed as a function of crank angle. It is used mainly for gasoline engines and is traditionally derived from cylinder pressure measurements [12].

Rate of heat release (RoHR). This is an absolute quantity, expressed as energy released (HR), or energy released per crank angle degree (RoHR), as a function of crank angle, that is, Joules or Joules/

degree. The cumulative heat release quantifies the combustion efficiency. It is traditionally derived from cylinder pressure measurements [13].

Combustion temperature. Unlike cylinder pressure, temperature is a spatially localized property in the combustion chamber (although an average temperature is sometimes computed from cylinder pressure). It is not measured with a temperature probe but indirectly via radiant emission, as realized by the two-color method in diesel engines [14], laser-induced fluorescence (LIF) [15], or coherent antistokes raman spectroscopy (CARS) [16] in gasoline engines. Knowledge of temperature distribution is useful since hot conditions give rise to NO_x formation (see below).

Kernel development. In gasoline engines, the kernel refers to the earliest moments of the flame. Minute differences in the growth and location of this kernel continue to influence the flame behavior long after it has left the spark-plug gap and are ultimately responsible for cycle-to-cycle variations [17].

Flame propagation. In gasoline engines knowledge of flame propagation is fundamental in optimizing engine performance. It reveals, for example, the effects of gas motion such as tumble or swirl. In flame propagation studies the moment at which the flame front arrives at a certain point is determined. The speed with which the flame must have travelled to reach that point can then be calculated and a vector established from which the spatial-temporal propagation of the flame is reconstructed. With an array of sensors the area of burned mixture existing at any particular moment is expressed as a polar diagram. Asymmetric flame-propagation profiles are generally undesirable.

2.2. SENSORS FOR NONOPTIMAL OR ABERRANT COMBUSTION

A major requirement in calibration is the need to avoid rough engine running at all costs, partly to meet customer expectations and particularly to avoid engine damage. To accommodate variations in production or variations with age or operating environment, conservative calibrations must be adopted to ensure that engines at the extreme ends of a distribution operate acceptably.

Cycle-to-cycle variations. The combustion process always varies from one cycle to the next in, for example, SOC, RoHR, and MFB. These variations arise from slight changes in air-fuel ratio and the random nature of turbulent flow fields. Normally cycle-to-cycle variations are slight, having been successfully minimized in the engine design, but field conditions can arise when they become excessive. The results are discernable to the driver as rough engine running.

Cylinder-to-cylinder variations. The sensors used in most engine management systems measure overall averages for all cylinders and cylinder-to-cylinder variations are not accommodated. One cylinder may receive slightly more air or fuel than another. For example, suppose that half the cylinders in an engine run rich and the other half lean. To an HEGO sensor mounted in the (combined) exhaust, this condition will be indistinguishable from all cylinders running at stoichiometry. Cylinder-to-cylinder variations can compromise the performance of the whole engine, since everything must be calibrated to suit the problem cylinder(s). This situation could be avoided by managing each cylinder separately.

Misfire. A misfiring engine releases unburned air-fuel mix or charge. This degrades drivability through rough engine running, damages catalytic converters, and causes excess hydrocarbon emission. Legislation demands on-vehicle detection of this condition. Detecting full and persistent misfire on a single cylinder is fairly easy. The problem is that misfire is not an either/or condition. Some charges may burn partially, with late, aberrant burning in the expansion or exhaust strokes. It is not uncommon for one or more cylinders to misfire occasionally. This intermediate state is more difficult to diagnose, since the temperature of the catalyst and the output from an HEGO sensor do not provide unambiguous indications. The favored method of misfire detection is via crank speed, that is, by detecting torque fluctuations. However, this diagnostic method is inadequate for operation at high engine speeds, for engines possessing a large number of cylinders [18], and on rough roads. This forces the implementation of long integration windows, making the control system slow, whereas entry into a misfire condition can be rapid, especially when operating close to the lean limit. The crankshaft method remains popular, however, because it is cheaper to implement than methods using combustion sensors.

Knock. Knock is an unwanted form of gasoline engine combustion caused by improper charge ignition ahead of the propagating flame. The propensity of an engine to knock is mainly due to excessively advanced ignition timing, or the use of a high compression ratio. Knock generates sonic pressure waves which reverberate within the cylinder. These are transmitted audibly to the surroundings and can damage the engine structure. Knock is characterized by its intensity. The measurements normally used to assess knock are the maximum peak-to-peak amplitude, or an RMS measurement of amplitude over time, and the onset time, or the ignition timing at which knock appears.

The usual antiknock strategy is immediate retardation of the ignition timing, as this gives the necessary speed of response [19]. For reasons of economy and performance, it is desirable to run an engine as close as possible to the knock condition. The problem is detecting the onset of knock. The customary but indirect method uses an accelerometer mounted on the engine block. However, there are problems with this approach, particularly at high engine speeds, when knock can be difficult to distinguish from background noise caused by mechanical phenomena such as piston slap and valve operation [20].

It is also necessary to identify an appropriate sensor location on the engine block. Acoustic nodes are obviously undesirable and the single accelerometer must be capable of monitoring all cylinders.

Combustion noise. The rate of increase in cylinder pressure, whether in a gasoline or diesel engine, is related to combustion noise. It is expressed as the rise rate $dp/d\theta$, where p is pressure and θ is crank angle. On diesel engines this quantity is a function of the amount of fuel injected during the ignition delay, that is, the premixed burn. Manufacturers seek to limit the rise rate to reduce noise. An alternative (but expensive) approach is to fit deadening or attenuating materials around the engine. A further possible approach is to retard the injection (on a diesel engine) or the spark (on a gasoline engine). There are likely to be practical difficulties with this method, however, since the ECU does not know $dp/d\theta$. The effects of vehicle environment (i.e., altitude and temperature) on rise rate are also difficult to predict.

Lean-burn limit. On gasoline engines the use of a lean air-fuel ratio offers better fuel economy than the more commonly used stoichiometric condition. With increased leaning, however, combustion becomes unreliable and increased cycle-to-cycle variation and frequent misfiring may occur. This is perceived by the driver as rough running and poor drivability. Combustion at the lean limit is very sensitive to small changes in air-fuel ratio. The flame propagates more slowly and the time taken by the flame front to traverse the combustion chamber varies. To obtain the best performance from lean-burn technology it is necessary to operate as close as possible to the lean-burn limit, but this limit is difficult to define. A UEGO sensor can be used to obtain the air-fuel ratio from the exhaust gas composition, but this information does not unambiguously indicate the condition of misfire.

Lean-burn engines were the subject of a great deal of work in the 1990s but have not, so far, been put into production. Improvements in the performance, cost, and availability of sensors, together with rising fuel costs, are likely to mean that interest in lean-burn technology returns.

Cold-starting. Fuel management during cold-starting is notoriously difficult. Partial fuel evaporation and the survival of some unburned fuel into subsequent cycles can cause erratic engine running. Considerable charge enrichment is necessary to ensure that sufficient fuel vaporizes to form an ignitable mixture. Small differences in fuel metering have profound implications for vehicle pollution (white smoke is characteristic of cold-starts). Temperature sensors located in the engine coolant or on the cylinder head are often slow to respond. On a gasoline engine no feedback signal is available from the HEGO sensor until normal operating temperature is reached; hence, fuel metering during start-up is normally open-loop.

Transient operation. Engines are normally well optimized for steady-state running, that is, when operating points are maintained for an extended period. Engine calibration maps are populated using just such tests. In practice, however, operation of a powertrain is characterized by frequent transient events where in-cylinder conditions vary from cycle to cycle. Out-of-engine sensors provide only slow feedback, generally averaged over a number of cycles. In traditional engine management transient maneuvers are approximated by a sequence of steady states, and this approach is frequently inadequate. Fuel quantity and timing can be adjusted rapidly in accordance with driver demand, while air mass flow, exhaust gas recirculation, and boost pressure respond much more slowly. Transient engine control is a very complex subject, which continues to be a topic of much research. Recurring problems include turbo lag, misfire, knock, and poor control of the air-fuel ratio [21].

2.3. COMBUSTION SENSORS FOR ENGINE INPUTS

Air-fuel ratio (AFR). On port-injection gasoline engines, the achievement of a stoichiometric AFR is essential to ensure maximum conversion efficiency in a three-way catalyst (TWC). Some residual liquid fuel always remains in the inlet manifold, so not all of the injected fuel enters the cylinder. This phenomenon is known as "hang up." During changes in operating point this hang up can be responsible for undesirable rich or lean spikes in AFR, compromising the efficiency of the TWC. On diesel engines AFR has not, historically, been a controlled variable, but this situation is changing because of after treatment (see below).

Fuel scheduling. Traditionally each combustion event required a single-injection event. However, modern designs make increasing use of multiple-injection strategies (pilot injection, main injection, post injection). For some operating points there may be as many as six injection events. Each event is characterized by two parameters: fuel quantity and timing. Aging and wear in the injector needle are accelerated by the increased number of events. The very small holes in modern injector nozzles are subject to tight manufacturing tolerances, with increased cost and manufacturing complexity. Fuel metering is easily affected by the gradual accumulation of nozzle deposits. With small pilot injections, say, 1–2 mg/stroke, a typical fueling error of around 0.5 mg/stroke is much greater by proportion than for the main injection of around 40 mg/stroke. On a diesel engine small errors in pilot injection can have large effects on combustion noise. Maintaining appropriate fuel metering in the absence of closed-loop control is very challenging.

Exhaust gas recirculation (EGR). Some inert exhaust gas is recycled and mixed with the engine input air to lower combustion temperatures and hence decreases the quantity of NO and NO2 gases (known as NO_x) present in the exhaust. In broad terms the more EGR is used, the "cleaner" the exhaust. However, there are obvious limits. For gasoline engines the constraints on EGR are combustion stability, cycle-to-cycle variation, and misfire. On a diesel engine the main constraint is the production of soot. EGR is normally monitored by measuring the opening of the EGR valve, and also by the reduced throughput of fresh air as indicated by the mass airflow (MAF) sensor.

Fuel composition. Four key parameters govern the relationship between fuel composition and combustion:

1. Evaporation characteristics, paramount in engine starting and warm-up
2. Ignitability, defined in diesel fuel by the Cetane number
3. Susceptibility to knock, controlled in gasoline via the octane number
4. Fuel blend

In the field, fuel composition is normally carefully controlled. There are cases where compositional variation becomes a factor. Specification of a fuel by its Cetane number alone may not be sufficient for nonstandard fuels or universally applicable to all engines [22]. The octane number of gasoline is affected by combustion chamber deposits in ways that are poorly understood [23] and varies more widely with natural gas than with gasoline [24]. It is also useful to be able to run vehicles on gasoline-alcohol blends [25].

Barometric pressure. Engine operation is inevitably affected by barometric pressure since, via density, this controls the available oxygen in the charge. Uncompensated high-altitude operation causes diesel engines to emit more smoke.

Ambient temperature. Air-charge temperature also determines the quantity of oxygen available and so affects engine operation in a similar manner to barometric pressure. Hot environments, if uncompensated, cause diesel engines to emit more smoke.

Air mass. The quantity of air available to an engine determines the quantity of fuel that may be injected [26]. The amount of air drawn into an engine depends upon the gas dynamics and volumetric efficiency of the air path. Only one mass airflow (MAF) sensor is normally fitted to an engine, so the transducer measures total airflow into all cylinders. Problems may arise due to air leaks or through exhaust gas remaining in the cylinder from the preceding cycle (residuals). Air estimation is made more complex if variable valve timing (VVT) is used, since current legislation demands on-vehicle verification. The measurement of valve timing is either performed by a hydraulic sensor or by measuring the phase difference between crankshaft and camshaft. Neither measurement gives direct confirmation of the effect of VVT on combustion.

Maximum brake torque (MBT). In spark-ignition engines MBT refers to an optimum spark timing for which a defined amount of fuel and air will deliver the greatest torque. This optimum timing is a compromise between early SOC, which wastes energy in opposing the piston, and late SOC, which fails to realize maximum cylinder pressure [27]. MBT depends on flame propagation and burn duration, and is usually determined via empirical relationships from cylinder pressure.

2.4. COMBUSTION SENSORS FOR ESTIMATING ENGINE OUTPUT

Before considering engine output sensors the following definitions are required.

Load. The load developed by an engine is expressed in several ways: as power, torque, indicated mean effective pressure (IMEP) and brake mean effective pressure (BMEP). On a conventional gasoline engine, power output is regulated by an air intake throttle. On a conventional diesel engine, power output is controlled by the quantity of fuel injected. Most modern engines make use of multiple injection strategies and/or variations in AFR to manage after-treatment (see below), and these variations affect power. No unique relationship exists between any one controlled parameter and load. For example, if five fuel injections per cycle are used, any number of fuel quantities and timings can generate the same output power. The ECU lacks any direct measure of engine load. Load is measured during engine calibration and estimated during operation from measurements of volumetric efficiency, engine speed, air mass, and inlet manifold pressure. However, this estimation process is vulnerable to any variation in ambient conditions.

Specific fuel consumption (SFC). SFC describes the conversion of fuel energy into useful work and, thereby, also expresses the fraction of fuel energy lost or wasted. SFC can be considered as the fuel flow required to generate a specified power output. The fuel economy of a vehicle is a similar parameter, which expresses the amount of fuel consumed per unit distance driven. Obviously, these parameters should be minimized. Emissions of CO_2, being simply an aspect of fuel burned, are legislated indirectly via fuel economy standards. This information is determined in the development stage; it is not verified directly by an ECU.

Exhaust temperature. After-treatment devices (see later) must be thermally managed for proper operation and this depends, in turn, on engine-out exhaust temperature. Currently this is assessed with out-of-engine sensors.

Exhaust oxygen. In exhaust after-treatment (see later), noxious emissions are reduced by oxidation in the exhaust system. Thus, oxygen availability in the exhaust is a critical factor. The amount of oxygen in the exhaust gas stream is available in principle via a UEGO sensor but not for each cylinder.

Emissions. The four emissions subject to statutory regulation are the oxides of nitrogen (NO_x), unburned hydrocarbons (HC), carbon monoxide (CO), and particulate matter (PM). Considerable efforts are made to restrict all four of these emissions in the engine exhaust. Almost all contemporary vehicles are fitted with after-treatment devices of various designs, which attempt to clean the exhaust gas. The technology of after-treatment is becoming very complicated. At the time of writing, particulate filters are about to be introduced on a widespread scale for diesel engines. In a few years, deNO_x catalysts will probably be required as well. Good exhaust gas management is essential if operation of these devices is not to be compromised. This management is achieved by elaborate injection schedules—four, five, or even six injections may be used within one engine cycle, depending on the region of the operating map.

Two relatively new forms of internal combustion engine offer potential applications for combustion sensors: the homogeneous-charge compression-ignition (HCCI) engine [28] and the direct-injection spark-ignition (DISI) engine [29]. The main drawback of the HCCI engine is the need for precise ignition control. If the fuel-air charge is excessively diluted, the combustion is incomplete or misfires. If the charge is insufficiently diluted, the combustion is violent and knock may occur. A need exists, therefore, for a combustion sensor that can detect the precise moment of auto-ignition.

The DISI engine uses a lean, inhomogeneous charge at low load. Potential uses of combustion sensors here are to maintain combustion stability at the lean limit, to verify that stratification of the charge is proceeding as intended, to help control engine-out emissions of NO_x and PM, and to monitor the multiple-injection schedules required to manage the complexities of after-treatment.

3. DESIGN ISSUES WITH COMBUSTION SENSORS

As noted earlier, this review focuses primarily on combustion sensors with potential for installation in production vehicles. Production vehicles are not, however, the only potential field of application:

Parallel needs arise on a regular basis in engine research and development where novel laboratory-grade sensors can prove to be of value. Many of the sensors discussed in this section were designed many years ago, and although serial applications are known, these were never extensive. This want of adoption can be ascribed to various causes, among which three are of particular importance.

First, cost-based business decisions must be taken. In many cases sensors required by new emissions legislation were not available at a suitable price, so other more economical means were found. Emissions legislation has had a "technology-forcing" action, however, and this is likely to continue.

A second reason is that, in an engine test cell, signal processing is not normally a problem. Should real-time processing prove impossible, then postprocessing is often an acceptable alternative. In the laboratory this processing is performed by additional hardware, whereas in a vehicle this may not be possible since the real-time signal processing capability of an ECU is limited.

Signals from combustion sensors, expressed as a function of crank angle, show various features. The signal normally peaks near TDC during the course of the burn and other complex features are common. There is thus no single, unique reading but rather a string of data points between the leading and trailing flanks of the peak. The features of most relevance are not always obvious and can require algorithmic extraction. This topic is not covered in any detail here, first, because the technical hurdles are more relevant to a book on digital signal processing and, second, because the solutions are often proprietary. The features of interest include signal levels, gradients, and integrals between certain points. The signal processing is often undertaken by a neural network [30]. In general analytical relationships linking signal features to the combustion process may not be available and this forces the adoption of empirical correlations.

Third, the engine compartment of an automobile is not a sensor-friendly environment and the combustion chamber can be markedly hostile. The pressures, and especially the temperatures, are high. In an engine test cell a life expectancy of a hundred hours, or a recalibration interval of a hundred hours, may be acceptable. In the field, vehicle manufacturers require operation for hundreds of millions of cycles with minimal (and preferably no) servicing requirements. These problems have so far precluded the widespread use of combustion sensors on production vehicles.

Ruggedness has an immediate bearing on packaging—an extremely important topic but one which seldom receives the attention it deserves. First, the sensor installation must penetrate the combustion chamber, and the most practical access point is not necessarily the best for sensing purposes. Engineers may not be able to cut access ports into the cylinder head. Space is almost always restricted, especially on multivalve engines. Cylinder heads and walls are highly stressed, and great precision is required if any machining is required. Finally, combustion chambers are surrounded by oil and cooling water circuits.

From the servicing standpoint, access must be provided to facilitate sensor cleaning or replacement. Choice and flexibility as to the installation are therefore greatly restricted. If a sensor is to be readily accepted for production vehicles, awkward or costly modifications to engine architecture should be avoided.

The best sensor installations are those that require no additional access. Integration of the sensor within another component is common, such as the glow-plug (on diesel engines), the spark plug (on gasoline engines), the fuel injector (on direct-injection engines), or the head gasket (on all engines). For laboratory investigations of diesel engines, a common practice is to replace the glow-plug with a sensor (as long as cold-starting is not required). If the bore has been opened, adaptor fittings are available, which allow glow-plug reinstallation when necessary. More relevantly to the present discussion, the installation of miniaturized combustion sensors within glow-plugs and spark plugs, without compromising the original function of either, is certainly possible. The two platforms, incidentally, are not equivalent. Heat rejection from glow-plugs is inherently poor, and this runs counter to the cooling requirements of many sensors.* Spark plugs are available with off-center electrodes, although these designs are admittedly expensive and may well differ in their thermal and electrical characteristics.

In considering packaging requirements, it should be noted than an ideal combustion sensor should be nonintrusive, or at least nonperturbing, inasmuch as any protrusion into the combustion chamber risks local distortion of the flow field. In-cylinder fluid flow is absolutely critical in the combustion process. Flow fields determine the propagation of the turbulent premixed flame in a port-injection gasoline engine and the air-fuel mixing of the turbulent diffusion flame in a diesel engine. For these reasons a flush-mounted, or perhaps recessed, sensor is clearly preferable.

Operating temperatures vary markedly with mounting location. The bulk temperature of the cylinder head during engine operation is unlikely to exceed 150 °C, but during a hot soak the absence of forced convection may allow somewhat hotter conditions to occur. A strong thermal gradient (say, 50 °C) exists across the cylinder head, from the cooler inlet valves to the hotter exhaust valves. A combustion chamber wall reaches 250 °C, but probe tips (thermally decoupled from the heat-sinking engine structure) become considerably hotter, as much as 650 °C [31]. Temperatures in the central region of the combustion chamber are hotter still and may exceed 2000 °C. Heat shields, cooling fins, and water jackets impede miniaturization, so thermal ruggedness is a property worth striving for in a sensor design.

Electronics and signal conditioning are equally important in packaging. The burden born by the contemporary ECU is advantageously lightened if some aspects of signal conditioning (e.g., amplification or temperature compensation) are integrated within the sensor unit. Electronic circuitry able to withstand high-temperature operation is expensive. To accommodate all loads and speeds, the signal processing unit must handle widely varying signal levels, and this can cause problems. For example, the gain may need decreasing if a high threshold is exceeded, or increasing if a low threshold is not exceeded. Current drawn from the vehicle battery, small for a single sensor, rapidly becomes inconvenient for multisensor installations. Last but not least, the electronic package must guard against earth loops and electromagnetic interference (EMI). Appropriate shielding is needed, and sensors of low output impedance are preferable.

The design problems engendered by temperature constraints, EMI, and earth loops highlight the advantages of optical signal transmission. For example, optical fibers can be run alongside high-voltage cables with impunity. This mode of signal transmission is obvious, and indeed inherent, in luminosity measurements, but other possibilities exist, as will be described. Manufacturing costs are admittedly inflated by the high precision needed in positioning and aligning the optical fibers—critical to achieve good optical sensor performance. This is not the place to discuss optoelectronics, but a potential solution proposed for a combustion sensor is fiber location via micromachined silicon fingers [3].

The final design aspect discussed here is sensor fouling. Sooting (i.e., carbon deposition) is not always an issue. Inherently suitable applications are those where the sooting propensity of the fuel is low (e.g., natural gas) and those where the sooting propensity of the engine is low (port-injection gasoline engines). DISI engines have a problem with soot generation when running in stratified mode.

---

* Little appreciated is the ability of the glow-plug heating element to act as a temperature sensor, and thereby detect misfire (Kong, H. 1991. Application of Glow Plugs for Combustion Sensing in a Diesel Engine. SAE 911878.).

Historically the diesel engine has had the largest difficulty in this area. Modern diesels may be regarded as low-sooting, since if the engine exhaust is clean, then so are the in-cylinder gases. Other fouling agents are possible. The historic problems caused by leaded fuel are obviously vanishing rapidly, but some lubricants contain inorganic elements such as phosphorous and calcium, which do deposit [32], and these may not be so easy to dislodge as carbon.

The in-cylinder environment is determined by duty cycle and sooty deposits are, to some extent, oxidizable, provided conditions are sufficiently hot. Highly loaded duty cycles, or at any rate mixed-mode cycles with intermittent high-load operation, are inherently favorable for soot removal from probe tips [33]. It should be recognized that, while surface accretions are often found within engine cylinders, the depth of these accretions has, as a rule, stabilized, so that on sensor tips the depth of the deposit may stabilize—although not necessarily before the signal is completely blocked. Further research is required in this area.

A modicum of self-decontamination or self-protection can be designed into transducer probe tips. This involves the exploitation of high temperatures for oxidation and violent shear forces for re-entrainment. Probe tips standing clear of the surrounding surface appear less vulnerable to soiling. The reasons are not well understood, but it may be that such deposits are poorly protected by the thermal boundary layer and, thereby, more effectively oxidized [34]. (Deposit friability is also likely to influence re-entrainment [33]. The probe tip geometry has a bearing on soiling, since angular profiles with corners apparently discourage deposition, unlike rounded profiles. Highly polished surface finishes are preferable to rough-hewn ones [35]. The periodic application of sparks to maintain tip cleanliness has not been reported, but this cleaning mechanism is suspected when sensors are installed in spark plugs [36]. Similarly, combustion sensors installed within glow-plugs benefit from periodic purging; indeed, the plugs might beneficially be operated specifically for this purpose, even when the engine is warm.

In-cylinder soot deposition has been comprehensively addressed for emissions control, and some of these ideas could be redeployed to advantage in sensor design. It is likely that soot is stored temporarily on the wall of the combustion chamber where it is shielded from burnup by a thermal boundary layer. During the expansion and exhaust strokes some of this soot is re-entrained and emitted. Should this emission mechanism be significant, efforts to control soot via the bulk gas would be misplaced. Modeling and experimental work [37] strongly implicate thermophoresis in the transferal of soot particles across the thermal boundary layer. In the reverse process, for example, the temperature of an optical probe was varied and the light attenuation used to measure the thickness of the attached deposit [37]. The deposition rate was faster with a cooler window. This seems to prove that the way to protect sensors is to raise tip temperatures: not so much to oxidize the soot once it has deposited but to weaken the thermophoretic force and, thus, discourage deposition in the first place.

4. TRANSDUCERS AND PRESSURE MEASUREMENT

Techniques for cylinder pressure measurement are well established [38]. Engine development would nowadays be unthinkable, indeed impossible, without it [39]. There is a good reason for this: pressure reveals, directly and unambiguously, the fundamental operation and purpose of the engine, namely, the force exerted by the combustion gases on the piston. Pressure measurement provides, therefore, the key measure of engine performance.

At the same time it is instructive to consider, albeit briefly, why in-cylinder pressure sensors are needed at all. Combustion (which is a controlled explosion) generates stresses and strains within the surrounding engine structure. The cylinder head bolts are, therefore, likely to carry the same information as any pressure sensor—and this data may be made available by strain gauges or load cells. Strain gauges may be preferable, if only because failure of a load cell can dramatically reduce the clamping force in a head bolt [40]. For either sensor, however, reverberations within the engine structure and mechanical operation of the valves degrade the signal-to-noise ratio (SNR). Thus for precision measurements, in-cylinder pressure sensors are unlikely to disappear. However, where less accurate assessments of SOC, for example, will suffice, head-bolt strain may well be adequate [41]. The fact that one such sensor is able to monitor more than one cylinder is an undoubted advantage.

4.1. DESIGN AND CONSTRUCTION

Commercially pressure sensors are a mature, mainstream technology [42]. However, as explained in section 3, these sensors are generally unsuited for production vehicle use. Thus research continues into alternative designs and operating principles. The current most successful approaches are based on piezoelectric or piezoresistive devices.

The most common cylinder pressure measurement device is undoubtedly the quartz piezoelectric transducer. Research continues into alternative piezoelectric materials, since ceramics of various (often proprietary) formulations offer manufacturing advantages over quartz [43]. Candidate ceramics are polycrystals of lead niobate ($LiNbO_3$) and lead titanate ($PbTiO_3$), and single crystals of lithium niobate ($LiNbO_3$), gallium phosphate ($GaPO_4$), and silicon dioxide (SiO_2) [44]. When selecting a piezoelectric material, thermal tolerance is the critical constraint, because, first, this determines the durability of the device and, second, the piezoelectric effect vanishes above a certain threshold (known as the Curie temperature). In material selection, this thermal tolerance is traded against another property, that of pyroelectricity, where charge is generated not mechanically but thermally through temperature fluctuations. Piezoelectricity and pyroelectricity are linear functions, respectively, of the first differential of force and the first differential of temperature. Pressure may be obtained by signal integration, provided the pyroelectric effect is small. Pyroelectricity can be reduced by adopting electrodes that run parallel to the axis of polarization [45]. A single-crystal piezoceramic will, in comparison to a polycrystalline material, generally offer greater thermal stability and lower levels of pyroelectricity. These advantages often compensate for lower sensitivity and greater manufacturing complexity.

Piezoresistive materials can also be used as pressure-sensing devices, although this application is less well documented. The cited advantages over piezoelectric materials are ease of signal processing, including temperature compensation, and better noise rejection, courtesy of the low output impedance [46]. These advantages arguably offset lower sensitivity in comparison with piezoelectric materials. A square-shaped silicon wafer is mounted such that the applied force, constant current, and measured potential are all mutually perpendicular [47]. Current is applied across two opposing corners and the potential measured across the two other corners [48]. The potential is a function of the load to which the wafer is subjected since the electrical conductivity is changed anisotropically. The distributions of stress and conductivity within the wafer are carefully optimized using finite element modeling. Other researchers report the use of piezoresistive bridges [49].

Since piezo devices have limited temperature tolerance (< 250–300 °C), they are normally protected within a sensor housing, where they receive, indirectly, the strain undergone by a metal diaphragm [46]. In the design described by Sugitani and others [48], figure 5.1a, the strain is transmitted via a ceramic force-transmission rod, a stainless-steel hemisphere, and a force-transmission block. The piezo is sandwiched between the transmission block and a pedestal, and the whole assembly stands within a hermetically sealed capsule. The force-transmission rod is ceramic to reduce axial heat transfer. The function of the hemisphere is to equalize and distribute the transmitted force. The pedestal and the transmission block are made from devitrified glass.

The integration of piezos into existing features is a well-researched topic. It is possible, for example, to install piezos within a functioning glow-plug [50] and within the washer of a spark plug, sandwiched between two electrodes [51]. In this latter approach, cylinder pressure causes relative movement between the central electrode and the body of the spark plug. Various designs are reported by Morris and Li-Chi [52] and Morris [43]. The piezoelectric properties are lost beyond a certain pressure, and in spark plug torque-down specifications this limit is easily exceeded. Spark plug resonant frequency is another design consideration [51]. The washer-shaped piezoelectric elements depicted in figure 5.1b–c are aimed at integration into fuel injectors and glow-plugs [53]. As with spark plugs, cylinder pressure forces the central region of the unit to move relative to the cylinder head. This movement compresses the piezo transducer. In the two designs depicted it is argued that the pin, rather than the housing of the glow-plug, represents a superior transmission medium for cylinder pressure.

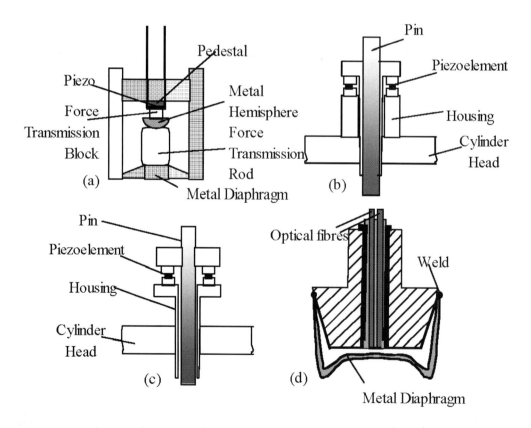

Figure 5.1. Various designs of pressure sensor

Using printed circuit technology it is possible to install an array of piezo devices within a cylinder-head gasket [54]. (The purpose of installing more than one pressure sensor is to study knock, as explained later). A resin packing is used to limit the mechanical stress introduced by the clamping force between engine block and cylinder head; and a copper film, crimped around the cylindrical face, acts as a fire wall and gas seal. A laminated design uses two spacer layers through which channels are cut for signal lines, and a modular design routes signal lines through laser-drilled tunnels [55].

Serious interfacing problems have to be overcome with piezoelectric devices [43]. The electrically noisy environment in the vicinity of an internal combustion engine is a problem for all piezo-based systems. A salient difficulty with the spark plug as a sensor location is that it is very susceptible to ignition noise, through electrostatic or electromagnetic induction [52].

Thus many researchers have abandoned development of piezo devices and look instead to optical methods of pressure sensing, for which more than one operating principle is certainly conceivable. The obvious application of optical fibers is for signal transmission. The merit of this approach is that it avoids the suffusion of heat energy into the sensor body, as unwittingly facilitated by a force-transmission rod. There are other options. Morris [43] proposed making optical fibers perform double duty, as sensor and transmission medium, by using the loss of total internal reflection consequent on fiber distortion when under load. He tested several design configurations in which these fibers were installed within spark plugs.

It is possible to monitor the deflection of a diaphragm optically, via a measurement of incident and returned light. The signal is nonmonotonic, since the deflection in the diaphragm changes not just the total distance travelled by the light but also the area of intersection between the incident and received light as projected onto the diaphragm [56]. But, provided the deflection is limited, a positive characteristic is observed. The pressure sensor depicted in figure 5.1d [57], and reported in several variant designs [58–61], measures the light reflected from a flexing hat-shaped metal diaphragm. Component dimensions are reportedly optimized to provide internal compensation for thermal expansion and contraction. The incident light is emitted by an infrared light-emitting diode (LED) and the returned light is measured by a silicon photodiode. The ability to detect small deflections is improved by tapering the fiber's end [62]. The design is sufficiently diminutive to allow installation into standard rather than off-center spark plugs. The sensor is self-monitoring insofar as the photodiode measures directly the output from the LED. Full-scale deflection of the diaphragm is 10–15 μm. The optoelectronic module is mounted onto the sensor. A critical aspect of the design is the hermetic seal, since the reflective properties of the diaphragm surface are modified by oxidation. (For a detailed description of the mechanical assembly see reference 61.) By a similar principle, Sasayama and others [56] report a silicon diaphragm of 300 μm thickness and 3 μm maximum displacement.

A further and intriguing design uses interferometry, or the phase difference between two interfering optical beams, to measure the deflection of a silicon diaphragm [63]. The interferometer, light source, and photodiode are all integrated within an optical signal-processing chip. Micromachined v-shaped grooves in the silicon-wafer substrate provide a positive location for the optical fibers with respect to the waveguide, also etched into the substrate, reducing cost and manufacturing complexity. The optical fiber terminates just short of a light-reflective diaphragm, the intervening distance being 50–100 μm, depending on deflection. Light is reflected from the fiber tip but also from the surface of the diaphragm. Because the light reflected from the diaphragm travels an additional distance across the gap, this causes destructive and constructive interference between the two returning paths. The spectrum is then analyzed by a path-imbalanced Mach-Zehnder interferometer. Since the necessary information is not encoded in the intensity of the light, the device is unaffected by age or bend-related degradation in light transmission.

4.2. PRINCIPLES AND CHARACTERISTICS

A pressure sensor may be either distally or proximally mounted. In distal mounting the sensor is connected to the cylinder via an adapter or flexible tube down which pressure waves propagate. The sensor experiences a lower operating temperature since it is thermally decoupled from the engine. Ease of access is also facilitated since the narrow flexible tube may be conveniently routed, for example, along the outside of a spark plug [43]. However, distal mounting is not without problems since the passageway may be blocked by soot. The signal is then attenuated and phase shifts are introduced due to the additional pressure-wave propagation distance. The intervening gas column, moreover, imposes its own organ-pipe resonances on the signal. The geometry of the tube is thus a crucial factor [64]. For these reasons installation guidelines discourage recessed-bore mountings, in which the probe tip falls short of the combustion-chamber wall, as these particularly encourage organ-pipe acoustics. These resonances are easily confused with knock [65]. Organ-pipe acoustics are avoided simply by adopting the flush-mount. That said, the relationship between bore geometry and bore acoustics is simple, and it is possible, through appropriate design measures, to suppress these resonances [50].

Thermal expansion and contraction (especially through direct flame contact with the diaphragm) causes baseline drift; there is thus an uncertain offset, which appears as a slowly changing DC component superimposed on the signal. This situation is exacerbated in piezoelectric sensors that measure changes in pressure rather than absolute pressure. In short-term or intra-cycle drift the sensor fails to recover from the thermal shock of the combustion event with sufficient rapidity: The pressure peak then appears asymmetric and the trailing flank traces out a false path. In medium-term or inter-cycle drift, the sensor is affected by more gradual thermal changes as the engine moves from one operating point to another [66].

Fortunately temperature changes in the piezoelectric material itself take place over a timescale of several hundred milliseconds, and this period substantially exceeds the combustion window. Compensation is accomplished through judicious selection of the low-frequency time constant in the charge amplifier [31]. In packaging the sensor, various methods are used to slow down or attenuate heat transfer, for example, interconnecting passageways, heat shields, water or air cooling [42]. The transducer is also thermally connected via the sensor housing to the cylinder head [31]. A heat shield prevents direct flame contact and, due to its heat capacity, slows heat transfer. Gaps in the shield, through which the pressure waves are supposed to propagate, can become blocked with soot. The added mass of the shield lowers the natural frequency of the sensor. The passageways have a quenching action, as heat dissipation bars the flame from narrow gaps. Lee and others [67] present an alternative approach based on the use of an empirical equation to compensate for short-term drift.

Offset determination is known as cylinder-pressure referencing, or pegging—an involved topic for which only cursory treatment can be given here. The task is preferably performed every cycle but every few cycles may be acceptable for some applications. Various methods have been devised, and Randolph [68] examines no less than nine. Examples are referencing to inlet-manifold pressure during induction, referencing to exhaust-manifold pressure during exhaust, and pressure calculations using fixed or variable polytropic indices. Many of these techniques have merit, but all generate slightly different results and none are without pitfalls [69].

The operation of valves and fuel injectors often generates high-frequency pressure signals. Rejecting these is relatively straightforward, since the signal can simply be low-pass filtered [53]. As a precautionary measure it is possible to design the sensor housing such that these high-frequency components are poorly transmitted, with improvement in SNR. On this basis, for example, it has been argued that the design of figure 5.1d is superior to that of figure 5.1c. Should valve or injector operation be a nuisance, a cylinder-head mount is preferable to a block-mount since the head gasket acts as a vibration damper.

Typical cylinder-pressure traces for a diesel engine are depicted in figure 5.2 [70]. The "motored" curve refers to a nonfired cycle; it will be noted that this trace is, in fact, asymmetric and peaks not at TDC (where cylinder volume is a minimum) but a few degrees earlier. The pressure peak should not, therefore, be taken to indicate TDC. The phase difference between peak pressure and TDC (caused by the slower rate of cooling, in comparison with the faster rate of compression) is known as the thermodynamic loss angle (TLA). Derived parameters like IMEP depend strongly on accurate knowledge of TLA, and several measurement methods are available [71]. The fired cycles follow this motored curve up until the moment of ignition, after which combustion raises the pressure. After completion of the burn, the fired cycles again return to the motored curve.

4.3. APPLICATIONS

4.3.1. Pressure Sensors Used for Combustion Measurement

Unlike many other sensors, the pressure sensor supplies a fundamental thermodynamic property for which analytical, rather than empirical, relationships, are available and from which widely used parameters like HR, MFB, and combustion temperature can be derived. This field is well researched and the

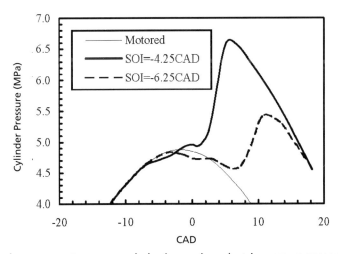

Figure 5.2. Cylinder pressure traces recorded using a piezoelectric pressure sensor mounted in a single-cylinder direct-injection diesel engine. Legend refers to one motored (nonfired) trace and two fired traces for which start of injection (SOI) is given in degrees before TDC, operating point, 1500 rpm/3 bar IMEP. One fired cycle in every four. Ensemble averages of 50 cycles.

pitfalls are well known. Problem areas are the polytropic index, ratio of specific heats, heat rejection, phasing with respect to TDC, signal noise, and pegging. The exact implications depend on the derived parameter; it should be particularly noted that small errors in pegging introduce large errors in derived parameters. To ameliorate the measurement errors introduced by these uncertainties, researchers must continue refining their thermodynamic models [72,73].

Empirical rules have been devised to relate maximum brake torque (MBT) to maximum cylinder pressure. It has been shown that maximum pressure follows MBT by a roughly constant interval [74]. The exact interval varies slightly from one engine model to another. On this basis the crank angle corresponding to maximum pressure may furnish the information necessary for closed-loop control of MBT. The spark is, in this form of control, retarded or advanced so that maximum pressure is maintained at, say, -12 °CA* [44]. Maximum cylinder pressure can be misleading if it arises solely through compression. Recognizing this, Herden and Küsell [49] determined MBT and SOC by mirroring the pre-TDC pressure trace about TDC. This mirrored image is then subtracted from the real pressure trace, post-TDC. The difference is integrated and the crank angle at which half this area had accumulated, then determined.

4.3.2. Pressure Sensors Used for Nonoptimal or Aberrant Combustion

Knock is manifested in cylinder pressure measurements in two ways: a sudden increase in gradient, part way up the leading flank; and (more obviously) a gradually decaying, high-frequency ripple on the trailing flank. Automatic detection of this condition is fairly simple. The ripple is filtered (isolated) from its base component, and its amplitude then compared to that of the expected (nonknock) noise background. The knock condition is flagged whenever the amplitude of the ripple exceeds a certain allowable threshold, expressed as a multiple of the background. Knock detection by this means is often more reliable than with block-mounted accelerometers [44]. However, even though the fundamental resonance of the combustion chamber and its harmonics are known from a theoretical standpoint, in practice there are various confounding factors. These include the precession of the piston, the position of the sensor with respect to nodes or antinodes, and variations in the speed of sound in accordance with changing thermodynamic properties. The pressure sensor has its own self-resonance, and this adds a further high-frequency component [75]. The appearance of frequencies unexplained by conventional acoustic theory is common [76]. Hence, knock detection continues to be avidly researched, for example, via frequency-domain analysis [77] or the thermodynamic properties of the end-gas [78].

Misfire appears as a lower than expected pressure, the lower limit being the motored curve. At low speeds, and especially at idle, pressure may not be raised sufficiently during normal combustion to furnish an unambiguous diagnostic [44]. Hence, rather than peak pressure, an integrated function may be better [49]. Partial burns late in the expansion stroke or in the exhaust stroke may be portrayed ambiguously as the signal approaches the noise level. The symmetry of the curve about TDC may, therefore, offer a better diagnostic approach. When cylinder-pressure traces for all cylinders are superposed, this method is analogous to detecting fluctuations in crank speed without inertial effects, the smoothing action of which may conceal crucial information [79]. Due to combustion instability the lean-burn limit is detected in a manner akin to that of misfire: The geometrical mean of the pressure traces for all cylinders is computed; the trace of each cylinder compared to this mean; and, if smaller, more fuel is injected on that cylinder [44].

---

* The convention adopted in this chapter is to specify crank angle with respect to TDC, positive signifying before TDC, negative after.

The management of combustion noise is straightforward insofar as dp/d is the industry-wide metric: Timing is simply retarded until this gradient falls below a predetermined threshold [44]. To manage engine roughness during cold-starting, the pressure sensor possesses an advantage insofar as its signal is, in principle, available from the moment at which cranking commences; and it could, therefore, indicate, via computation of HR, the AFR during this critical period [80].

4.3.3. Pressure Sensors for Engine Inputs

A pressure sensor measures (via the ideal gas law) the amount of air trapped within a cylinder, taking into account the exhaust-gas residuals and vaporized fuel ingested during the intake stroke [81]. Deformation and vibration in the cylinder head and the action of cam on valve spring introduces discontinuities into the signal [82]. This allows detection of valve operation.

In VVT these events are referred to crank angle, and the valve timing may be optimized. Compensation for variations in the evaporation characteristics of fuel, through changes to injector timing and fuel quantity, is performed through an analysis of cycle-to-cycle variations and estimation of the power theoretically available but not delivered. Closed-loop control of exhaust gas recirculation (EGR) is accomplished using the same method as for the lean-burn limit described earlier [44].

4.3.4. Engine Output Pressure Sensors

Engine load is available through IMEP as this parameter is defined in pressure terms. From this, torque is estimated. Engine-out NO_X is available from cylinder pressure, since the dominant parameter in NO_X formation is combustion temperature which is derived from cylinder pressure. Hence, if maximum cylinder pressure is above a certain threshold, the timing is retarded [44].

5. OPTICAL SENSORS AND LUMINOSITY

Light-emitting hydrocarbon flames are generated within an internal combustion engine. The light produced is measured as luminosity, which is defined as the amount of light passing through a certain area or within a certain solid angle. Engine researchers use luminosity measurements in two ways. First, laser irradiation of the in-cylinder gases (often called laser imaging or laser diagnostics, see Zhao and Ladommatos [83]) has been used since the early 1990s to study combustion. The apparatus used for this is complex, and specially designed engines are required, which are often far removed from production models. Although laser imaging represents a topic of undoubted importance to the motor industry, it lies outside the remit of the present review.

A much more common use for optical sensors is the use of optical probes to view events within the combustion chamber. These investigations normally use production engines and are, therefore, reasonably near market.

There are overlaps between the two fields. For example, it is possible to introduce a laser beam through one optical fiber and measure the returned luminosity via a second. The fraction of light successfully traversing the gap (more correctly, the fraction lost through absorption) has been shown to be related to the AFR of the intervening gas [84]. Optical probes are sometimes used for spectral evaluation of the luminosity. Certain chemical species betray their presence by radiating at characteristic wavelengths [85]. However, most studies simply measure luminosity and ignore the spectral information.

Even if the exact source of the luminosity is not identified, the signal gives useful information with which to control and/or optimize the combustion process by making empirical correlations.

5.1. DESIGN AND CONSTRUCTION

The practical construction of an optical probe is relatively straightforward. An optical fiber is terminated behind a protective window of quartz, sapphire, or fused silica. Quartz will withstand temperatures up to 1100 °C. The greater cost of sapphire is offset by its ability to withstand temperatures up to 1500 °C [86]. Removal of significant sections of cylinder heads and so forth to accommodate these windows is not required, since optical probes can be made very small. Installation of the probe within existing features is the norm, with spark plugs being a common platform. The sparking operation is not necessarily compromised by the presence of an optical probe.

Two designs of optical sensor, one dedicated and the other installed within a spark plug, are shown in figure 5.3a–b [35]. In the spark plug mount, the center electrode is bored, and a small section of the obtruding outer electrode has been removed to clear the field of view [36]. Alternatively, the housing or shell may be bored and the center electrode encircled by a sensor array [87].

The glow-plug is less commonly reported as a sensor location. Although optical probes could probably be incorporated within a glow-plug, glow-plugs tend to be replaced with dedicated sensing units [88].

Optical sensors have also been installed within cylinder head gaskets, and as many as 150 channels have been reported [89].

The signal from an optical sensor is, in effect, a spatial integration from a defined volume of combustion chamber gas. This integration must be considered carefully, since if the volume is too large spatial resolution will be poor and local reaction zones or flame fronts may not be distinguished. If the volume is too small, signal levels are low. All optical sensors have an "angular acceptance profile," since light is collected from within a "cone of reception" controlled by the shape of the sensor tip. A flat-tipped device collects light from within an angle of around 20°, whereas a spherical or lens-shaped tip collects up to 40% of the light at an angle of 65° [90]. Microlenses can limit the acceptance cone to as little as 4° [89].

The orientation of the reception angle with respect to the propagating flame affects the relationship between luminosity and other combustion parameters since it fixes the instant at which the flame crosses the sensor field of view. In principle asymmetric probe-tips could be used to correct for nonideal sensor entry angles. Probes must be keyed to allow repeatable alignment. Alternatively, two probes with intersecting fields of view can be used to reconstruct the flame path [91]. Four probe-tip designs, and an acceptance profile for one of them, are given in figure 5.3(c)–(d) [35].

In a significant (albeit complex) departure from the traditional approach reported by Ikeda and others in 2005 [92], light is collected and focused by two mirrors, the first of which is concave, the second convex. This apparatus is similar to an optical telescope and has been sufficiently miniaturized to allow its incorporation within a commercial spark plug. Light is focused by two aluminum-coated surfaces for which the focal length is 3 mm. The measured volume, 0.8 mm long by 0.1 mm wide, lies at a distance of 5.5 mm from the spark gap [93]. This measured volume is considerably smaller than can be achieved by traditional probes.

Although traditional optical probes are relatively straightforward to use, the associated signal-conditioning apparatus is not always convenient. Commonly used laboratory apparatus includes spectrographs, image-intensifiers, photocathodes, monochromators, photomultiplier tubes (PMT), optical

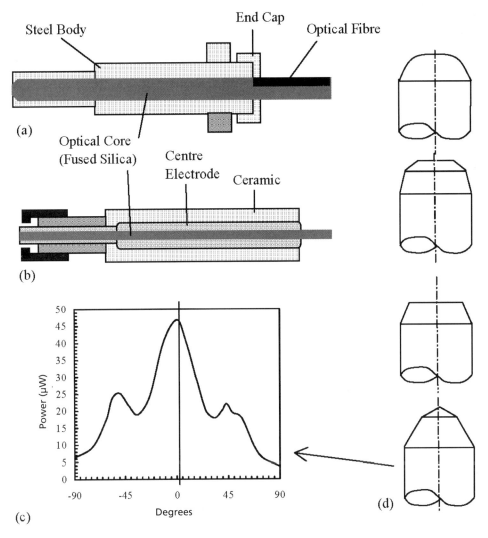

Figure 5.3. Optical probe designs: (a) a dedicated probe; (b) installation in a spark plug; (c) light acceptance profile of a probe tip; (d) four probe-tip profiles

filters, and charge-coupled devices (CCD) [94]. Mercury-vapor lamps are used for calibration [92], and diffraction gratings of different resolutions and bandwidths are required to examine spectral features. Less accurate measurements are possible with lower-priced equipment such as semiconductor photodetectors. The need for precision alignment with optical fibers is relaxed when large-area photodiodes are used [36]. Optical-fiber strands can be separated to permit the parallel use of several photodetectors, each receiving differently filtered light [95]. The advantage of conducting simultaneous multiwavelength measurements is that this allows signal ratios to be related to AFR as discussed later.

The combination of window (and probe tip geometry), optical fiber, and signal conditioning used determines the accessible spectral range, since each component imposes its own transfer function on the signal. For example, a photomultiplier may effectively exclude wavelengths of > 650 nm, although some attenuation is often observable at 400 nm [96]. A signal is of limited use without knowledge of the intervening transmission characteristics or transfer functions.

5.2. PRINCIPLES AND CHARACTERISTICS

Luminosity measurements cover three distinct regions of the electromagnetic spectrum: ultraviolet (1–400 nm), visible light (400–700 nm), and infrared (700 nm–100 μm). Chemi-luminescence (light emission through chemical reactions) is characteristic of the visible and ultraviolet. Molecular vibrations give rise to infrared emissions. Many chemical species emit at characteristic wavelengths as shown in table 5.1. There is also a broadband, pseudo-blackbody emission from the red end of the visible range well into the infrared. This is caused by incandescent soot.

Luminosity measurements are qualitative rather than quantitative, since some uncertainty always surrounds the depth of view. Light emitted from a point must pass through some intervening gas before it reaches the detector, and the absorbing properties of that gas will change the signal. The location of the measurement point with respect to that of the propagating flame must also be considered. In a gasoline engine a sensor located at the ignition point will view the rear of the flame as it recedes. A sensor located within the end-gas region views the front of the same flame as it advances. While some chemical species are directly created by and radiate from within the flame, others are heated indirectly and radiate from outside the flame. Chemi-luminescence in the visible and ultraviolet regions arises from within the flame, whereas infrared emission is normally due to burned gas. Some infrared emission (known as the "adiabatic foot") is apparent prior to ignition and is due to compression heating [33]. Similarly, cooling gas in the burned zone may continue to radiate in the infrared long after the passage of the flame. This is termed "afterglow." Useful information does not usually emanate from the

Table 5.1. Characteristic emission wavelengths of several important combustion intermediates or products [32,86,97]

Species	Wavelength (nm)
OH	306.0
CH	314.0, 387.1, 431.5
HCO	318.6, 329.8
HCHO	395.2
C_2	473.0, 516.5, 564.0
H_2O	691.0, 710.0, 927.7, 1800, 2700
CO_2	2800, 4300

same region of the electromagnetic spectrum throughout the combustion process. An infrared sensor may, for example, respond slowly if preflame reactions radiate at wavelengths outside its range [40].

The accumulation of light-attenuating soot on probe tips is an important problem for optical sensors; indeed, this is their chief drawback. Antisooting techniques were described earlier in this chapter. In certain instances signal degradation has been found to be due to deterioration of the optical properties of the sensor window itself, for example, through devitrification of quartz [90]. The attenuation of light from this cause is wavelength-dependent and mainly occurs in the visible region of the spectrum. The absence of any consistent relationship between this attenuation and running time may indicate randomly repeating cycles of deposition and re-entrainment [98]. A recessed tip provides some protection, perhaps by arresting thermophoretic deposition. However, too much recession reduces the sensor's field of view.

In automatic control of combustion using optical devices, algorithms based on detecting the crossing of an absolute signal level should be avoided since signal levels are often reduced by the accretion of probe-tip deposits. While trigger thresholds can be reduced to some extent, they cannot be placed too close to the noise band. Probe-tip deposits are always manifested as a drop in gain, and this causes feedback errors leading to artificial shifts in set point. To circumvent these problems, gain-independent criteria must be used in which signals are normalized according to their respective peaks [33]. Noise rejection is improved by referencing the signal to a nonlighted, black condition, provided the infrared emission from the hot wall of the combustion chamber is taken into account[*] [34].

In gasoline engines luminosity sensors sometimes (but not always) detect the spark [94]. This can be a nuisance since designing algorithms to reject spark-related features is not trivial [99]. This phenomenon does, however, confirm proper operation of the spark plug; and the spark discharge, just like the combustion it initiates, is thus open to investigation via emissions spectroscopy [100].

In general chemi-luminescence in port-injection gasoline engines gives rise to a distinct peak of the form shown in figure 5.4 [30]. The leading flank of this peak contains, or sometimes merges with, a much smaller peak [96] or a foot [99]. It has been speculated that the flame propagation is briefly arrested so that the flame front hovers within the vicinity of the window [99], or it may be that the first peak corresponds to the flame itself as it passes through the field of view. The second peak may be due to burned gas lying behind the flame and subjected to continued heating and compression. According to Kawahara and others [93], CH, C_2, and OH all appear within the flame. The first two of these are said to disappear rapidly and thus act as markers for the flame proper. Active formation of the third component (OH) continues for some time within the burned gas, leading to a second peak.

With HCCI engines there is no flame front. Two clearly demarcated peaks are apparent within the chemi-luminescence. The first is small and relates to the so-called cool flame. The second peak is much larger and corresponds to the main burn. Although these two peaks differ in the intensity of luminosity, their spectral contents are similar [101].

In diesel engines infrared emission (due to soot incandescence) is the main feature of interest. The luminosity shows two clearly defined features: first, a small peak relating to the premixed burn; and, second, a rapidly following dominant peak relating to the mixing-controlled burn [33]. The transition between the two phases is easily pinpointed. It should, however, be noted that some workers report soot incandescence midway through the premixed phase [70]. Early features attributed to precombustion

---

[*] Intriguingly, if one were to block the optical tip with some blackening compound, then this would make a temperature sensor, courtesy of the blackbody radiation. This idea does not seem to have been exploited.

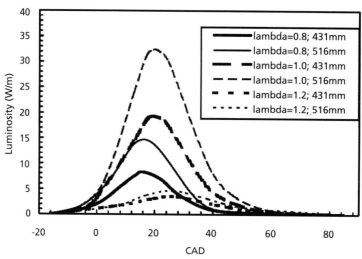

Figure 5.4. Luminosity as measured in a port-injection gasoline engine. Lambda is air-fuel ratio (AFR) normalized to stoichiometry. Legend refers also to wavelength in the electromagnetic spectrum. Operating point, 2500 rpm/60 kPa. Ensemble averages of 100 cycles [30].

reactions before the emergence of the flame proper have also been reported [102]. In DISI engines luminosity reveals soot associated with stratified operation [89].

5.3. APPLICATIONS

5.3.1. General

The first appearance of luminosity, while defining SOC to some extent, is strictly defined as the moment that light first enters the sensor field of view. In gasoline engines flame propagation may actually have been in progress before this moment—this explains phase differences between luminosity and other in-cylinder measurements. Difficulties in defining SOC may also be caused by the leading foot [99]. The ratio of luminosity measured at two wavelengths shows a sharp peak at SOC and may be less ambiguous [35]. The logical inference from this observation is that at SOC luminosity changes more strongly as a function of wavelength than as a function of crank angle. In diesel engines the strongest upturn in the signal heralds not SOC but the transition from the premixed phase to the mixing-controlled phase.

Chemi-luminescence [103] is closely related to pressure [36,96] and hence to parameters normally derived from pressure such as heat release (HR) [35,96,104] and mass fraction burned [96]. This association is particularly close for HCCI engines [105]. These observations lend credence to an analytical expression relating, for port-injection gasoline engines, luminosity to the mass and temperature of the burned gas [106]. The curves thereby derived were advanced by up to 2 °CA relative to measured luminosity. However, the model was nevertheless found to be a good predictor. Following further development it was possible, using luminosity measurements, to make reasonable estimates of HR.

The direct computation of local combustion temperatures from luminosity should not be confused with average combustion temperatures derived from pressure measurements. In gasoline engines flame temperature can be derived from OH [107]. In diesel engines OH is drowned by wideband soot incandescence, but this incandescence also allows temperature to be estimated. This estimation is classical pyrometry in which emissivity is a function of surface temperature as represented by Planck's law. To a first approximation, emissivity is independent of wavelength and the soot is uniformly distributed [108]. The method involves measurement of emissivity at two wavelengths [14].

Spatio-temporal measurements of luminosity in gasoline engines show the fastest flame propagation near stoichiometric AFR. As the mixture is leaned the flame front becomes thicker and moves more slowly and less predictably [93]. For the same crank angle, the area of burned gas enlarges with increasing load [95]. As might be expected from the flow field, flame propagation, although proceeding outward in all directions from the spark plug, is biased toward the exhaust valves. In DISI engines the elapsed time between spark and flame is several times longer in stratified mode than in homogeneous mode [102].

5.3.2. Nonoptimal or Aberrant Combustion

During partial misfiring, peak luminosity becomes variable and the signal often fragments into a sequence of confused spikes [95]. During complete misfiring the signal often disappears completely [34]. The lean limit is evident in cycles where the flame fails to reach the end-gas before the exhaust valve opens [36]. Chemi-luminescence at various wavelengths, or ratios thereof, can indicate cycle-to-cycle variability [92].

The occurrence of knock is indicated by gradually decaying oscillations on the signal's trailing flank. Reverberating pressure waves heat the burned gases, causing them to repeatedly reilluminate. An alternative view is that the density of emitters is briefly increased by compression. Incipient or borderline knock may increase the gradient of the leading flank without necessarily imposing oscillations on the trailing flank [35]. It is possible, using an array of luminosity sensors, to construct the direction taken by the pressure waves and thereby deduce the area of auto-ignition within a combustion chamber [87]. The transition from normal combustion to knock is particularly apparent in HCO and OH [109]. In normal combustion a single peak was observed, which decays slowly over about 20 °CA. On the appearance of knock this peak is replaced by spikes, the sharpness and fragmentation of which depend on the intensity of the knock. For weak knocking the principal luminosity peak is preceded by a smaller rise lasting for around 5 °CA. This is probably due to a cool (blue) flame presaging the main flame.

5.3.3. Engine Inputs

Luminosity-based closed-loop control systems are well documented. Hartman and others [88] use measurements of late-phase luminosity in a diesel engine to compensate for variations in barometric pressure and air-charge temperature. The fuel quantity was adjusted to maintain the same signal level at a prescribed late crank angle. The same method was also able to compensate for minor variations in injector characteristic affecting delivered fuel quantity. Infrared emission of the mixing-controlled burn reveals ignition delay and Cetane number [110]. Using this method an engine management system was shown to maintain the correct SOC regardless of variations in Cetane number and barometric pressure via automatic adjustment of injection timing (SOI) [111]. A similar approach has been used for closed-loop control on gasoline engines, for example, for spark timing [35]. Luminosity has been shown to be well correlated with AFR [112], and, as depicted in figure 5.4, stoichiometry generates the highest peaks [30]. Signal ratios between two species (viz., C_2/CH, C_2/OH and OH/CH) are strongly related to AFR [92]. This property is probably due to the AFR-related green or blue bias in the flame [30].

5.3.4. Engine Outputs

In diesel engines, where wideband soot incandescence dominates luminosity, the potential sensor application is obvious. Soot is generated predominantly in the mixing-controlled phase, rather than in the earlier premixed phase. Luminosity simply follows the rise and fall of in-cylinder soot, as soot formation gradually gives way to soot oxidation. Following end of injection (EOI) especially, the trailing flank of luminosity reveals partly the amount of late-burning soot and partly the loss of incandescence arising from gradual cooling. However, some luminosity is still apparent, even at exhaust value opening (EVO). This late-phase signal indicates that some soot survives until EVO and is thereby emitted from the engine. Using this principle a diesel engine can be controlled such that it continues to emit the same quantity of soot irrespective of variations in barometric pressure and air-charge temperature [88]. A similar control system can control EGR, since the amount of in-cylinder soot generated is a function of the EGR rate [94]. In port-injection gasoline engines, the appearance of infrared radiation, rather than just chemi-luminescence, suggests improper combustion via a soot-forming diffusion flame of unvaporized fuel [32]. From luminosity it is also possible to use the well-known Zeldovich equations [113] to calculate the concentration of in-cylinder NO_x [107].

6. ELECTRICAL SENSORS AND ION CURRENT

Hydrocarbon flames generate ions (see, e.g., reference 114). The electrical aspects these ions bring to the combustion process are well known [115]. These ions are mainly positive and are caused by the ionization of molecules. However, capture by molecules of free electrons also creates some negative ions. When an electrical field is applied, the ions migrate according to the potential gradient, and this migration gives rise to a small ionic current. This current may be used to reveal useful features about the combustion: this is the principle of ion sensing.

6.1. DESIGN AND CONSTRUCTION

The most practical realization of an ion sensor, at least in gasoline engines, is by use of the spark plug. This secondary sensing role should not to be confused with the primary one, since the ion current relates not to the spark but to the immediate aftermath of the spark. However, the spark itself (or rather the plasma created by the spark) can also be used for monitoring purposes, since the shape of the voltage pulse (especially the peak amplitude and decay time) reveals whether the air-fuel mixture is normal, contains insufficient fuel, or indeed is flammable. The voltage pulse has been shown to lengthen as the mixture is leaned, and also on the introduction of EGR [116]. Diagnosis of cycle-to-cycle variability may also be possible [117].

The electrical circuit used for ion sensing is conceptually simple and is shown in figure 5.5a [118]. A potential difference and current-sensing resistor are applied to the electrode gap as shown in figure 5.5b. Four important design choices must be considered:

1. The first question is what constitutes a suitable potential and resistance. A voltage between around 30–500 V and a resistance of a few hundred kΩ are typical.
    It is possible that the continual removal of charge from the ion cloud to support the measured current may distort the measurement. Low signal levels are occasionally seen (especially in HCCI engines), and there is, presumably, a limit to the voltage that may be applied. Experimentally, it has been shown that the measured current depends on voltage raised to the power of 0.8 [119].

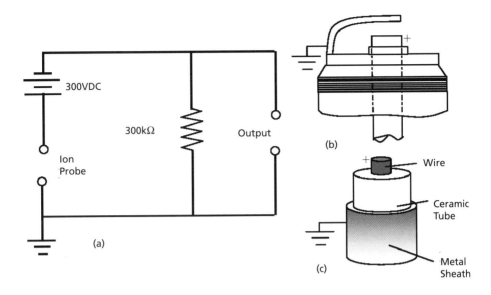

Figure 5.5. Electrical circuit (a) used to measure ion current; (b) as realized in a spark plug; or (c) a simple one-wire construction

2. A second question is the provision of a suitable earth connection, which, customarily, is the outer electrode of the spark plug. This includes the cylinder head, the cylinder walls, and the piston. In multisensor installations the sensors can have a common ground without compromising SNR. However, proper shielding and isolation of the signal conduit to avoid crosstalk is essential [118].

3. A third question is polarity or bias. Normally the center electrode is positively biased with respect to the outer [120], since negative biases have been found to yield poor currents [121]. Precisely how bias affects the ion cloud is uncertain. It is likely that the relative mobilities of electrons and ions are involved [122]. Free electrons (which have greater mobility) are rapidly collected by the positive electrode. This leaves flames with a net positive charge and a tendency for movement toward the negative electrode. This effect is called the Chattock electric wind [123].

4. The fourth question concerns the geometry and configuration of the electrode. The ion current depends on the area from which ions are collected. This is not surprising since small details in the electrode configuration are critical to the ignition ability of the plug [124]. Since the electrodes will, if too large, arrest flame development by acting as heat sinks [125], there are likely to be similar thermal effects on the ion current. The electrode gap is usually around 0.6 mm. Larger gaps increase the signal but only up to about 1.1 mm [126].

The often-cited advantage of ion sensing—that no additional sensor is required—is to some extent negated by the practical difficulty of packaging the current-sensing resistors, since the ignition system itself has to be modified. This is a manufacturing problem not faced by laboratory researchers using purpose-built equipment. For a traditional, distributor-based ignition system, integration within the ignition cables or distributor cap is necessary. For a distributorless ignition system, the current-sensing resistors are normally housed within the ignition coil [127]. These issues are not trivial, since expensive redesign of the tooling used for winding ignition coils is necessary. The bias voltage is introduced to the secondary winding of the ignition coil [120]. Should ion sensing become more widely adopted, these design constraints will become an important issue for ignition system development [128].

The ion-sensing function has a further effect on ignition-circuit design, namely coil ringing and dump-down. The electrical disturbance introduced by the ignition process easily encroaches on, and may dominate, the ion-sensor signal. There may be difficulty in separating the two signals. Ignition noise is not easily filtered out when of comparable frequency to the ion signal [118]. The introduction of noise from ignition events in other cylinders is also hard to remove. Given the unlikelihood of eliminating this disturbance, the best solution is probably to ensure its complete disappearance prior to arrival of the ion signal. This may require a nonstandard, one-coil-per-cylinder ignition system with which to apply a shorter pulse of greater intensity [129]. Considerable variations do arise in the voltages and pulse lengths necessary to achieve reliable ignition [130]. At high engine speeds the ignition pulse duration must not exceed 0.5 ms, although longer periods are possible at low engine speed. The circuitry designed by Lee and Pyko [127] disables ion sensing during ignition and waits until completion of coil discharge. With the ignition system described by Förster and others [126], spark duration may be adjustable independently of spark energy. This gives more flexibility in the choice of ion-sensing window. Further details of suitable electronic circuit designs for use with ion sensors are available from Collings and others [131] and Nicholson and Witze [132], while Miyata and others [122] describe methods whereby the signal can be isolated from the ignition pulse. Short-duration sparks are also an aspect of spark plug design [133].

Although in ion-sensing spark plugs not infrequently serve as both sensor and igniter, this double duty is purely for convenience. It is not essential and, where appropriate, the igniter function can be discarded [134]. In the last decade a large literature has been generated on ion sensing for HCCI engines where the igniter function is superfluous, giving more freedom of choice in plug location. An array of ion-sensing spark plugs has been mounted around the cylinder head in one reported study [135]. This also means that the plug configuration may be modified in order to investigate ion-sensing principles. Removal of the side electrode, for example, increases the signal strength. This may be related to shielding of the gap by the side electrode and consequent limitation of the volume of gas subjected to the ion current [136].

Glow-plugs offer a similarly convenient platform [137] and may be used for ion sensing simply by modifying their electronics. The great versatility of ion sensors is most evident when spark plugs are dispensed with completely. In its simplest form an ion sensor consists of just one wire, referenced to ground, and insulated from ground by a surrounding ceramic tube [138]. Successful designs include copper wire insulated with an epoxy coating, and titanium wire insulated with a specially oxidized layer [139]. If necessary, the ceramic tube can be protected from breakage by placing it within a protective metal sheath as shown in figure 5.5c. Extremely narrow probes are possible with this mode of construction, and several can be inserted within a spark plug shell encircling the central electrode. Alternatively, wires or metal strips can be installed within the cylinder-head gasket [118]. Gaskets lend themselves well to printed circuit board technology since multilayered wires can be routed to several cylinders. Successful trial designs include glass-fiber reinforced with polyimide resin and sandwiched between steel or Kevlar [132], and glass epoxy laminate electroplated with nickel and gold sandwiched between Teflon [140]. Low durability through thermal degradation of the supporting gasket material was reported, however.

6.2. PRINCIPLES AND CHARACTERISTICS

An investigation of the ion species giving rise to ion current has been made [141]. Although conventionally labeled "ion current," this label is technically incorrect since free electrons (which have a higher charge mobility than free ions) presumably play some role [137]. The three most common ion-formation mechanisms are summarized below:

$$CH + O \rightarrow CHO^+ + e^- \quad (5.1)$$

$$CHO^+ + H_2O \rightarrow CO + H_3O^+ + e^- \quad (5.2)$$

$$H_3O^+ + e^- \rightarrow H_2O + H \quad (5.3)$$

The product of reaction 1 is CHO^+. This process is not likely to be dominant since CHO^+ ions are rapidly destroyed via recombination with water as shown by reaction 2. Reaction 2 shows that water is ionized in turn to form H_3O^+—the ion that probably attains the greater abundance. H_3O^+ ions are in turn reduced through electron recombination as shown by reaction 3 [142].

This explanation comes from ion-current studies using steady-state laboratory flames, and caution should be used in applying these results to the rapidly changing and unsteady conditions encountered within the combustion chamber of an engine. Detailed modeling suggests that NO^+ is the most abundant ion and the major source of free electrons later on in the burned gas zone, this being a consequence of the low ionization energy of NO [143]. The above description is greatly simplified, and the emergence of other species such as $C_3H_3^+$, CH_3^+, and CH_3O are likely. In summary, therefore, the identities of the main charge carriers are uncertain. However, this uncertainty has not prevented the use of ion sensors as practical investigative tools. Ion sensing has a long history of use in gasoline engines, particularly for flame-propagation studies [121].

Ion sensors appear to be quite resilient to electrode soiling. However, caution should be exercised since any electrically conducting deposits will permit current leakage. Ion-current decay is apparently protracted for a soiled probe [119]. Operation of the spark plug is, on occasion, also compromised by fouling [144]. Current leakage measurements show that plug fouling by soot and/or water condensation takes place following a cold start [145]. Gasket-mounted sensors (which lack the protective action of a nearby spark) show decreased signal levels after only an hour or so in use [132].

Ion sensors are found to be prone to soot deposition in diesel engines where conducting tracks introduce offsets to the signal [138]. The conductivity of soot deposits is not constant and appears to change during the course of a cycle. This may be due to compression of the soot by overlying gas. Short circuits can appear and then disappear as soot is re-entrained. When sooting is particularly prevalent, the ion current may be completely overwhelmed. For probe designs where one wire is insulated by a ceramic tube, the tube geometry determines the susceptibility of the transducer to short circuiting. Finally, soot particles can themselves carry charge [146], although the implications of this phenomenon for ion sensing have not been explored.

Fuel composition also has an effect. Nowadays the effects are unlikely to be due to lead, since leaded gasoline has largely been removed from use. Lead has, however, been shown to affect the ion cloud, possibly by interfering with reaction (3) [121]. Metal-containing compounds at trace levels are present in both fuel and oil. They may be there through inadvertent contamination or may have been added deliberately (in which case the chemical formulations are proprietary). These compounds are broken down in the combustion chamber and the metals released. These metals are not inert participants in the combustion process. For example, they can have either soot-promoting or soot-suppressing effects [147]. The precise mechanisms involved are not relevant: the important point is that these metals do ionize with implications for the ion flux [148]. This probably explains the effect of the manganese-containing gasoline additive Methylcyclopentadienyl Manganese Tricarbonyl (MMT) on the ion current. However, the oxygenating compound methyl tert-butyl ether (MTBE) also affects ion current, although the mechanism for this is unclear [126]. It is possible that interference with the ion flux takes place through organic as well as inorganic reaction pathways. Should ion current prove to be a function of the brand of gasoline, or of the time of year at which gasoline is purchased [119], then this is obviously a serious drawback. However, it appears that signal levels, rather than signal features or patterns, are distorted by this effect.

A typical ion-current trace for a port-injection gasoline engine is given in figure 5.6 [129]. Three distinct features are evident: an initial spike, followed by two peaks. The initial spike (which is probably due to the spark) can be ignored. The following explanations for the two peaks are reasonably well accepted. The first peak relates to the flame kernel created close to the electrodes. The second peak is due to burned gas remaining after the flame has propagated away from the electrodes. Different ion-formation mechanisms are at work in each peak. In the first peak, chemical reactions between neutral species generate sufficient energy to ionize the products; this is known as chemi-ionization. In the second peak, energy generated at the flame front (and probably also the ongoing compression) continues to heat the burned gas behind the flame. This causes thermal ionization, and so the second peak is sometimes called the post-flame peak.

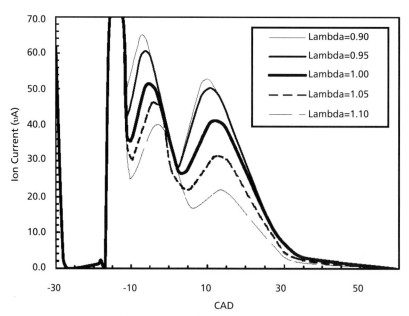

Figure 5.6. Ion current measured in a four-cylinder port-injection gasoline engine. Nonstandard one-coil-per-cylinder ignition system. Lambda is air-fuel ratio (AFR) normalized to stoichiometry. Operating point was 2500 rpm/85 nm.

In HCCI engines the ion current shows just one peak, probably caused by chemi-ionization. Combustion is markedly cooler in an HCCI engine than is the case for conventional engines. The single peak may also occur because combustion is distributed and occurs almost simultaneously throughout the combustion chamber (i.e., there is no propagating flame). The mixture is also diluted with greater amounts of exhaust gas or air in HCCI, which probably also explains the weaker ion probe signals of a few milliamps at best, with no usable signal at all for some operating conditions. Some workers consider ion sensors unsuited for use with HCCI engines.

In diesel engines a consistent pattern does not emerge. The signal characteristics are variable and depend on the operating point and on the location of the sensor with respect to the combusting plume [138]. From a theoretical standpoint, a clear demarcation between chemi-ionization (in the flame) and thermal ionization (in the burned gas) is unhelpful. The duality of diesel engine combustion (i.e., premixed burn and mixing-controlled burn) is evident in the signal. There may be two separate peaks or two closely coupled peaks, or the two peaks may merge. On occasion the merging is so complete that the premixed burn is only shown by a slight change in gradient on the leading flank of the peak. This wide variation in signal characteristics may indicate movement of the combusting plume as it grows. The signal may decline when the flame leaves the local area and then grow again as the plume leaves the combustion chamber and a greater volume of charge ignites.

6.3. APPLICATIONS

6.3.1. Combustion

The most obvious use of ion current sensing is to indicate SOC, as shown by the leading flank of the first peak or at least the first upgoing pulse. On gasoline engines this application is restricted to sensor locations in the immediate vicinity of the spark. For more distant locations the additional flame-propagation distance must be taken into account. The same up-going pulse indicates SOC in HCCI engines for which no propagation distance is expected. For diesel engines the limitations described in the previous section should also be considered. Even if the first up-going pulse does not correspond precisely to SOC, it may still represent some convenient surrogate event from which SOC can be deduced and thereby placed under closed-loop control [137].

Relationships between ion current and thermodynamic properties are predominantly empirical and perhaps nonlinear [149]. In port-injection gasoline engines an association of some sort has been observed between maximum cylinder pressure and the second ion-current peak [150]. After manipulation the pressure-based expression for MFB has been shown to correspond closely to ion current [151]. This indicates that thermal ionization is likely to be responsible for the second peak, since temperature is closely linked to pressure. Saitzkoff and others [143] formulated expressions for ion current in terms of combustion temperature, pressure, and MFB. However, the ability to predict pressure from ion current has been shown to be problematic. This may be due to high gas flows or turbulence in the vicinity of the electrodes [152].

Despite these deficiencies in theoretical understanding, ion sensors are commonly used in port-injection gasoline engines for flame-propagation studies. Sensor arrays may be mounted around the cylinder bore to sense the later stages of combustion, or around the spark plug to measure its inception. The ideal behavior shown in figure 5.6 is not displayed on every occasion [118]. Around the bore a confused sequence of up- and down-going spikes can appear. The first up-going peak may represent flame arrival, and the subsequent behavior perhaps reflects the complex and turbulent nature of the flame front. The flame-arrival time at the bore is poorly correlated with MFB [140]. The more distant the sensor is from the spark, the more variable the flame-arrival time. Even around the bore flame-arrival times (and their standard deviations) vary markedly from one location to another. Distant (from the spark plug) locations are therefore likely to be less useful for closed-loop spark control [132].

Ion-current studies indicate that flames do not propagate radially from the spark plug to form a hemisphere of burned gas. The flame's progress is strongly affected by local variations in turbulence, temperature, or AFR. For this reason ion sensors are very useful for optimizing spark plug location and flow fields [118].

In HCCI engines large variations in flame-arrival time of as much as 2 °CA between locations around the bore have been found [135]. These variations suggest that in an HCCI engine combustion does not occur simultaneously at all points. This may be due to local variations in temperature, AFR, or quenched combustion in close proximity to the cylinder walls, which act as a heat sink.

6.3.2. Nonoptimal or Aberrant Combustion

Ion probe traces normally indicate knock through oscillations on the trailing flank of the second peak. These oscillations correspond to natural resonances of the combustion chamber and are usually also seen in cylinder pressure measurements. At some operating points ion sensors reveal knock undetectable by block-mounted accelerometers. This includes inaudible knock at low engine speeds, and audible, heavy knock at high engine speeds [153]. Ion sensors are particularly suited to knock detection at high speed and low load, because strong signals are generated in this region. The second harmonic is also more readily distinguished with ion current probes than with cylinder pressure transducers.

Misfire is straightforward to detect with an ion probe as there is no signal. Detection of the intermediate condition of partial burn is also straightforward, since although ions are still generated, they appear at the wrong moment. Partial burn during the expansion stroke may be distinguished from partial burn in the exhaust stroke [153]. The idle and low speed/low load condition is a difficult operating region in which to diagnose misfire because signal levels are low [127]. Any period allowed for ion annihilation will also complicate the signal analysis. Particular problems may be experienced when ion detection is required outside the engine (such as in the exhaust port) where the sensor must await EVO [134]. Weak signals may be caused if few ions are generated. Alternatively, weak signals can also be the result of ion annihilation. The time available for recombination is also smaller at high engine speeds [154].

Variation in the arrival time of the ion pulse is a consequence of cycle-to-cycle variability. This observation can be used to establish the lean limit [155]. The procedure is as follows. A criterion is set such that the flame arrives within a certain crank angle range for a defined number of cycles. A few early arrivals are permitted, so that the cylinder is not too far from the lean limit, and a few late arrivals are also allowed. The fueling is then adjusted to maintain this condition. In principal a similar approach could be used to control cycle-to-cycle variations in HCCI engines [156]. However, the large cycle-to-cycle variations observed in ion-current traces may also result from the sensor itself rather than from the combustion process [157]. This issue awaits further research.

6.6.3. Engine Inputs

An important potential application for ion sensors is in AFR control. This application is obvious for port-injection gasoline engines, as shown by figure 5.6, but can also be used for HCCI engines [135]. The signal has, however, been shown to weaken or even disappear for the leanest charges. The best AFR-related aspect of the ion signal, according to Reinmann and others [129], is the magnitude of the first peak. Peak detection algorithms are, however, made difficult and can be complicated by fragmentation of the first peak and the need to decouple the initial spark-induced spike. Reinmann et al. used a computed average within a certain range of crank angle. It should be noted that figure 5.6 displays global AFR. However, the sensor measures local AFR. This may lead to further applications for ion sensing, since local variations in AFR at the spark plug are a major factor in the cycle-to-cycle variability of port-injection gasoline engines.

In port-injection gasoline engines the reduction in signal level arising from EGR offers further possibilities for closed-loop control. A similar application exists for diesel engines. However, the increased amount of in-cylinder soot caused by EGR may cause sensor electrode fouling and short-circuiting. In diesel engines it is possible to use ion current to control the quantity of fuel in the pilot injection, and thereby control ignition delay in the main injection [138].

6.3.4. Engine Outputs

A good correlation exists between the second peak of ion current and IMEP [158]. The signal is in general able to follow the lower work performed with a lean AFR [159]. The higher-exhaust hydrocarbon concentration consequent upon partial misfire (late ion current) and complete misfire (no ion current) is another obvious association [160].

7. CONCLUSIONS

The two main conclusions emerging from this chapter are as follows:

- Engine management currently depends on open-loop control of the combustion process, or closed-loop control via out-of-engine sensors.
- Direct measurement of the combustion process from in-cylinder sensors would provide considerable benefits.

If in-cylinder sensing was available, the combustion process would be better optimized for perturbing factors that are at present either poorly compensated or uncompensated.

7.1. POTENTIAL APPLICATIONS

Table 5.2 lists potential applications for combustion sensors. The fundamental parameters used are start of combustion (SOC), cylinder pressure, combustion temperature, heat release (HR), and flame propagation. Nonoptimal or aberrant combustion includes cycle-to-cycle variation, cylinder-to-cylinder variation, misfire, and knock. In these cases the performance of the engine may be compromised because of the need to accommodate one "nuisance" cylinder. Boundary conditions leading to aberrant combustion are often difficult to define. The inputs to an engine are air and fuel. The amount of available air is uncertain because of engine breathing issues and ambient conditions. The fuel quantity injected is vulnerable to injector aging. Engine outputs are exhaust temperature, emissions, load, and economy. The first two are dealt with by after-treatment. Finally, DISI and HCCI engines represent two new areas of application for combustion sensors.

Table 5.2 Applications to which combustion sensors of various forms have been put, as reported in the literature

	MECHANICAL (PRESSURE)	OPTICAL (LUMINOSITY)	ELECTRICAL (ION CURRENT)
PERTAINING TO COMBUSTION GENERALLY			
Start of Combustion (SOC)	49, 161	95, 99, 111	137, 138, 156, 157
Cylinder Pressure	Not applicable	36, 96, 97, 143	141, 143, 149–152, 156, 157, 159, 162, 163
Mass Fraction Burned (MFB)	39, 164	96, 97, 106, 150	143, 151, 157, 159
Heat Release (HR) or Rate of Heat Release (RoHR)	38, 39, 72, 73, 80, 165	95, 98, 101, 105, 106, 166, 167	149, 156, 167, 168
Combustion Temperature	165	106–108, 165, 169	143
Kernel Development and Flame Propagation	139	89, 92, 96, 98, 99, 102, 121, 167, 170–173	52, 118, 120, 125, 139, 140, 157, 167, 172, 183
PERTAINING TO NONOPTIMAL OR ABERRANT COMBUSTION			
Cycle-to-cycle Variability	38, 39, 117, 174	92, 101, 103, 117, 169, 170, 172	140, 156, 157, 172, 174
Cylinder-to-cylinder Variations	38, 82	Not found	Not found
Misfire	44, 49, 58, 79, 82, 125, 175	32, 34, 36, 96, 103	120, 125–127, 133, 150, 153, 160, 176–178
Knock	39, 44, 49, 51, 54, 59, 77, 78, 82, 164, 175, 179, 181, 182	32, 87, 89, 96, 109, 169, 173, 177	120, 126, 131, 150, 153, 176, 180, 183, 184
Combustion Noise	38, 44	Not found	Not found

TABLE 5.2 *CONTINUED*

Lean-burn Limit	44, 185	34, 155	177
Cold Starting	44, 80, 82	Not found	186
Transient Operation	38, 187	Not found	138, 160, 178
PERTAINING TO ENGINE INPUTS			
Air-fuel Ratio (AFR)	82, 164, 188–190	30, 32, 92, 97, 169, 170, 191	129, 150, 156, 163, 178, 186
Fuel Scheduling	38	111, 192	137, 138
Exhaust Gas Recirculation (EGR)	44, 81, 82, 175	Not found	137, 138, 160
Fuel Composition	Not found	111	186
Barometric Pressure	Not found	5, 111	
Ambient Temperature	Not found	88	168
Air Mass	38, 175, 188	Not found	168
Maximum Brake Torque (MBT)	44, 51, 82, 164, 175, 190, 193, 194	95	193
PERTAINING TO ENGINE OUTPUTS			
Load	44, 175	97	159, 163
Economy and Efficiency	Not found	Not found	Not found
Exhaust Temperature	Not found	Not found	Not found
Exhaust Oxygen	Not found	Not found	Not found
Emissions	44	32, 88, 89, 192	160

Note: The table is cross-referenced against cited literature. Emphasis is given to sensors of immediate practical use, and investigations relevant, or potentially relevant, to real-time or closed-loop control of the combustion process. Thermodynamic studies of combustion of a theoretical nature, in which the sensors are of subsidiary interest (for pressure especially, in relation to MFB, HR, etc.) are not reported.

ACKNOWLEDGMENT

I wish to acknowledge the technical assistance of my colleague Khizer Tufail in preparing some sections of this review.

ABBREVIATIONS

AFR	Air-fuel ratio
BMEP	Brake mean effective pressure
CA	Crank angle
CAD	Crank angle degrees
CCD	Charge-coupled device
CARS	Coherent antistokes raman spectroscopy
CO	Carbon monoxide
DISI	Direct-injection spark-ignition
ECU	Engine control unit
ECM	Engine control module
EGO	Exhaust gas oxygen (sensor)
EOI	End of injection
EMI	Electromagnetic interference
EGR	Exhaust gas recirculation
EVO	Exhaust valve opening
HC	Hydrocarbons
HCCI	Homogeneous-charge compression-ignition
HEGO	Heated exhaust gas oxygen (sensor)
HR	Heat release
IC	Internal combustion
IMEP	Indicated mean effective pressure
LIF	Laser-induced fluorescence
LED	Light-emitting diode
MAF	Mass airflow sensor
MBT	Maximum brake torque
MFB	Mass fraction burned
MMT	Methylcyclopentadienyl Manganese Tricarbonyl
MTBE	Methyl tert-butyl ether
PM	Particulate matter
PMT	Photomultiplier tube
NO_x	Oxides of nitrogen
RoHR	Heat release
SOC	Start of combustion
SOI	Start of injection
SNR	Signal-to-noise ratio
TDC	Top dead center
TLA	Thermodynamic loss angle

TWC   Three-way catalyst
UEGO   Universal exhaust gas oxygen (sensor)
VVT   Variable valve timing

REFERENCES

1. Stobart, R. K. 1997. The demands of cylinder event control. SAE paper 970617.
2. Palma, G., O. Scognamiglio, and M. Lavorgna. 2004. Low cost virtual pressure sensor. SAE paper 2004-01-1367.
3. Challen, B. J., and R. K. Stobart. 1998. Some more diesel engine sensors. SAE paper 980167.
4. Grunde, T., and S. Wallman. 1977. Development of the Volvo Lambda-Sond system. SAE paper 770295.
5. Yamada, T., H. Nobuhiro, K. Yoshide, and K. Takeishi. 1992. Universal air-fuel ratio heated exhaust gas oxygen sensor and further applications. SAE paper 920234.
6. Hasei, M., T. Ono, Y. Gao, Y. Yan, and A. Kunimoto. 2000. Sensing performance for low NOx in exhausts with NOx sensor based on mixed potential. SAE paper 2000-01-1203.
7. Baranzahi, A., A. L. Spetz, P. Tobias, I. Lundström, P. Mårtensson, M. Glavmo, A. Göras, J. Nytomt, P. Salomonsson, and H. Larsson. 1997. Fast responding air/fuel sensor for individual cylinder combustion monitoring. SAE paper 972940.
8. Assanis, D. N., F. A. Friedmann, K. L. Wiese, M. J. Zaluzec, and J. M. Rigsbee. 1990. A prototype thin-film thermocouple for transient heat transfer measurements in ceramic-coated combustion chambers. SAE paper 900691.
9. Boggs, D., and G. Borman. 1991. Calculation of heat flux integral length scales from spatially-resolved surface temperature measurements in an engine. SAE paper 910721.
10. Guezennec, Y. G. and P. Gyan. 1999. A novel approach to real-time estimation of the individual cylinder combustion pressure for S.I. engine control. SAE paper 1999-01-0209.
11. See, e.g., Larsson and Schagerberg 2004. SI Engine Cylinder Pressure Estimation Using Torque Sensors. Paper presented at SAE 2004 world congress. SAE Report no. 2004-01-1369.
12. Heywood, J. B. 1998. *Internal combustion engine fundamentals.* New York: McGraw-Hill.
13. Ibid., 508–9.
14. Zhao, H., and N. Ladommatos. 1998. Optical diagnostics for soot and temperature measurement in diesel engines. *Progress in Energy and Combustion Science* 24: 221–55.
15. Einecke, S., C. Schulz, V. Sick, A. Dreizler, R. Schieβl, and U. Maas. 1998. Two-dimensional temperature measurements in an SI engine using two-line tracer LIF. SAE paper 982468.
16. Choi, I., K. M. Chun, C-W. Park, and J. W. Hahn. 2000. End-gas temperature measurements in a DOHC spark-ignition engine using CARS. SAE paper 2000-01-0237.
17. Keck, J. C., J. B. Heywood, and G. Noske. 1987. Early flame development and burning rates in spark ignition engines and their cyclic variability. SAE paper 870164.
18. Klenk, M., W. Moser, W. Mueller, and W. Wimmer. 1993. Misfire detection by evaluating crankshaft speed: A means to comply with OBDII. SAE paper 930399.
19. Wagner, J., J. Keane, R. Koseluk, and W. Whitlock. 1988. Engine knock detection: Products, tools, and emerging research. SAE paper 980522.
20. Hirako, O., N. Murakami, and K. Akishino. 1988. Influence of valve noise on knock detection in spark ignition engines. SAE paper 880084.
21. See, e.g., Pilley, P. D., A. D. Noble, C. D. de Boer, and A. J. Beaumont. 1990. Dynamic engine modelling for improved transient control. Technical paper published by I. Mech. E. Combustion Engines Group, 11/7/90.
22. Johnson, R. T., and K. R. Schmid. 1988. Summary results, CRC Project CM-126, Cetane engine combustion. SAE paper 881616.
23. Barnes, J. R., and T. Stephenson. 1996. Influence of combustion chamber deposits on vehicle performance and tailpipe emissions. SAE paper 962027.

24. Soylu, S., and J. van Gerpen. 1997. Determination of knock sensor location on a heavy-duty natural gas engine. SAE paper 971705.
25. Leshner, M. D., C. A. Luegno, and F. Calandra. 1980. Brazilian experience with self-adjusting fuel system for variable alcohol-gasoline blends. SAE paper 800265.
26. Ohyama, Y. 2002. Air/fuel ratio and residual gas fraction control using physical models for high boost engines with variable spark timing. SAE paper 2002-01-0481.
27. Heywood, J. B. 1998. *Internal combustion engine fundamentals*, pp. 19, 375. New York: McGraw-Hill.
28. Epping, K., S. Aceves, R. Bechtold, and J. Dec. 2002. The potential of HCCI combustion for high efficiency and low emissions. SAE paper 2002-01-1923.
29. Zhao, F., D. L. Harrington, and M-C. Lai. 2002. *Automotive gasoline direct-injection engines*. Warrendale, PA: SAE.
30. Hunicz, J., D. Piernikarski, and A. Niewczas. 2004. Transient in-cylinder AFR management based on optical emissions. SAE paper 2004-01-0516.
31. Anastasia, C. M., and G. W. Pestana. 1987. A cylinder pressure sensor for closed loop engine control. SAE paper 870288.
32. Ohyama, Y. 1990. Study on mixture formation and ignition process in spark ignition engine using optical combustion sensor. SAE paper 901712.
33. Day, E., J. A. Kimberley, and B. J. Mehallick. 1989. Start of combustion sensor. SAE paper 890484.
34. Piernikarski, D. and J. Hunicz. 2000. Investigation of misfire nature using optical combustion sensor in a SI automotive engine. SAE paper 2000-01-0549.
35. Nutton, D., and R. A. Pinnock. 1990. Closed loop ignition and fueling control using optical combustion sensors. SAE paper 900486.
36. Sun, Z., P. L. Blackshear, and D. B. Kittelson. 1996. Spark ignition engine knock detection using in-cylinder optical probes. SAE paper 962103.
37. Suhre, B. R., and D. E. Foster. 1992. In-cylinder soot deposition rates due to thermophoresis in a direct injection diesel engine. SAE paper 921629.
38. Randolph, A. L. 1994. Cylinder-pressure-based combustion analysis in race engines. SAE paper 942487.
39. Amann, C. A. 1985. Cylinder pressure measurement and its use in engine research. SAE paper 852067.
40. Challen, B. J. 1987. Some diesel engine sensors. SAE paper 871628.
41. Mobley, C. 2000. Wavelet analysis of non-intrusive pressure transducer traces. SAE paper 2000-01-0931.
42. Soltis, D. A. 2005. Evaluation of cylinder pressure transducer accuracy based upon mounting style, heat shields, and watercooling. SAE paper 2005-01-3750.
43. Morris, J. E. 1987. Intra-cylinder combustion pressure sensing. SAE paper 870816.
44. Shimasaki, Y., M. Kobayashi, H. Sakamoto, M. Ueno, M. Hasegawa, S. Yamaguchi, and T. Suzuki. 2004. Study on engine management system using in-cylinder pressure sensor integrated with spark plug. SAE paper 2004-01-0519.
45. Kusakabe, H., T. Okauchi, and M. Takigawa. 1992. A cylinder pressure sensor for internal combustion engine. SAE paper 920701.
46. Takeuchi, M., T. Kouji, N. Yutaka, O. Yoshiteru, and C. Yoshiki. 1993. A combustion pressure sensor utilizing silicon piezoresistive effect. SAE paper 930351.
47. Nonomura, Y., K. Tsukada, T. Morikawa, M. Takeuchi, A. Hosono, and M. Kosugi. 1994. New piezoresistive force detector for automotive use. *Institution of Mechanical Engineers*, C481/014.
48. Sugitani, N., M. Ueda, M. Kosugi, and K. Tsukada. 1993. Combustion pressure sensor for Toyota lean burn engine control. SAE paper 930882.
49. Herden, W. and M. Küsell. 1994. A new combustion pressure sensor for advanced engine management. SAE paper 940379.
50. Sonntag, R., S. Brechbühl, M. Schnepf, and P. Wolfer. 2002. New ways to measure pressure inside diesel engines without the use of individual bore holes. In *Thiesel 2002 Conference*, eds. C. Arcoumanis, J. M. Desantes, and F. Payri. Berlin: Springer Verlag.
51. Randall, K. W. and J. D. Powell. 1979. A cylinder pressure sensor for spark advance control and knock detection. SAE paper 790139.

52. Morris, J. E., and Li-Chi. 1985. Improved intra-cylinder combustion pressure sensor. SAE paper 850374.
53. Moriwaki, J., H. Murai, and A. Kameshima. 2003. Glow plug with combustion pressure sensor. SAE paper 2003-010707.
54. Mazoyer, T., P. Fayet, M. Castagné, and J-P. Dumas. 2003. Development of a multi-sensors head gasket for knock localization. SAE paper 2003-01-1117.
55. Vialard, D. J. 2006. Seamless integration of combustion pressure sensors into a multi-layer steel cylinder head gasket. SAE paper 2006-01-1211.
56. Sasayama, T., S. Suzuki, M. Amano, N. Kurihara, S. Sakamoto, and S. Suda. 1987. An advanced engine control system using combustion pressure sensors. *Institution of Mechanical Engineers* 201, D1:55–59.
57. Ulrich, O., R. Wlodarczyk, and M. T. Wlodarczyk. 2001. High-accuracy low-cost cylinder pressure sensor for advanced engine controls. SAE paper 2001-01-0991.
58. He, G., A. Patania, M. Kluzner, D. Vokovich, V. Astrakhan, T. Wall, M. T. Wlodarczyk. 1993b. Low-cost spark plug-integrated fiber optic sensor for combustion pressure monitoring. SAE paper 930853.
59. He, G., and M. T. Wlodarczyk. 1994. Evaluation of a spark-plug-integrated fibre-optic combustion pressure. sensor. SAE paper 940381.
60. Poorman, T., S. Kalishnikov, M. T. Wlodarczyk, A. Daire, W. Goeke, R. Kropp, and P. Kamat. 1995. Multi-channel and portable fiber optic combustion pressure sensor system. SAE paper 952084.
61. Poorman, T. J., L. Xia, and M. T. Wlodarczyk. 1997. Ignition system-embedded fiber-optic combustion pressure sensor for engine control and monitoring. SAE paper (1998) 970845.
62. He, G., M. T. Wlodarczyk, and E. L. Moore. 1993a. Tapered fibre-based diaphragm-type pressure sensor. *SPIE* 2070:39–46.
63. Fitzpatrick, M., R. Pechstedt, and Y. Lu. 2000. A new design of optical in-cylinder pressure sensor for automotive applications. SAE paper 2000-01-0539.
64. de S Vianna, J. N., J. P. Damion, and M. A. M. Carvalho. 1994. Contribution to the study of measurement of dynamic pressure in the interiors of combustion engine cylinders. SAE paper 942401.
65. Wagner et al. 1998. Detection of multiple resonances in noise. *International Journal of Electronics and Communications* 5: 1–4
66. Henein, N. A., S. L. Marek, and A. Fragoulis. 1992. Error analysis of time-dependent frictional torque in reciprocating engines: Effect of cylinder gas pressure. *Tribology Transactions* 35: 516–22.
67. Lee, S., C. Bae, R. Prucka, G. Fernandes, Z. S. Filipi, and D. N. Assanis. 2005. Quantification of thermal shock in a piezoelectric pressure transducer. SAE paper 2005-01-2092.
68. Randolph, A. L. 1990. Methods of processing cylinder-pressure transducer signals to maximize data accuracy. SAE paper 900170.
69. Brunt, M. F. J. and C. R. Pond. 1997. Evaluation of techniques for absolute cylinder pressure correction. SAE paper 970036.
70. Kook, S., C. Bae, P. C. Miles, D. Choi, and L. M. Pickett. 2005. The influence of charge dilution and injection timing on low-temperature diesel combustion and emissions. SAE paper 2005-01-3837.
71. Kim, K. S. and S. S. Kim. 1989. Measurement of dynamic TDC in SI engines using microwave sensor, proximity probe and pressure transducers. SAE paper 891823.
72. Eriksson, L. 1998, Requirements for and a systematic method for identifying heat-release model parameters. SAE paper 980626.
73. Brunt, M. F. J., and K. C. Platts. 1999. Calculation of heat release in direct injection diesel engines. SAE paper 1999-01-0187.
74. Heywood, J. B. 1998. *Internal combustion engine fundamentals*. New York: McGraw-Hill. Page 375.
75. Brunt, M. F. J., C. R. Pond, and J. Biundo. 1998. Gasoline engine knock analysis using cylinder pressure data. SAE paper 980896.
76. Dues, S. M., J. M. Adams, and G. A. Shinkle. 1990. Combustion knock sensing: Sensor selection and application issues. SAE paper 900488.
77. Scholl, D., C. Davis, S. Russ, and T. Barash. 1998. The volume acoustic modes of spark-ignited internal combustion chambers. SAE paper 980893.

78. Worret, R., S. Bernardt, F. Schwartz, and U. Spicher. 2002. Application of different cylinder pressure based knock detection methods in spark ignition engines. SAE paper 2002-01-166.8.
79. Komachiya, M., N. Kurihara, A. Kodama, T. Sakaguchi, T. Fumino, and S. Watanabe. 1998. A method of misfire detection by superposing outputs of combustion sensors. SAE paper 982588.
80. Leisenring, K. C., G. Rizzoni, and B. Samimy. 1995. Methods for internal combustion engine feedback control during cold-start. SAE paper 950842.
81. Hart, M., M. Ziegler, and O. Loffeld. 1998. Adaptive estimation of cylinder air mass using the combustion pressure. SAE paper 980791.
82. Sellnau, M. C., F. A. Matekunas, P. A. Battison, C-F Chang, and D. R. Lancaster. 2000. Cylinder pressure-based engine control using pressure-ratio-management and low-cost non-intrusive cylinder pressure sensors. SAE paper 2000-01-0932.
83. Zhao, H. and N. Ladommatos. 2000. *Engine combustion and diagnostics.* Warrendale, PA: SAE.
84. Nishiyama, A., N. Kawahara, and E. Tomita. 2003. In-situ fuel concentration measurement near spark plug by 3.392 μm infrared absorption method: Application to spark ignition engine. SAE paper 2003-01-1109.
85. Gaydon, A. G. 1957. *The spectroscopy of flames.* New York: John Wiley and Sons.
86. Pendlebury, M. A., and C. O. Nwagboso. 1994. Fiber optic sensors for engine combustion intensity detection: A review. ISATA 94EN060.
87. Töpfer, G., J. Reissing, H-J. Weimar, and U. Spicher. 2000. Optical investigations of knocking location on S.I.-engines with direct-injection. SAE paper 2000-01-0252.
88. Hartman, P. G., S. L. Plee, and J. E. Bennethum. 1991. Diesel smoke measurement and control using an in-cylinder optical sensor. SAE paper 910723.
89. Philipp, H., G. K. Fraidl, P. Kapus, and E. Winklhofer. 1997. Flame visualisation in standard SI-engines: Results of a tomographic combustion analysis. SAE paper 970870.
90. Antoni, C., and N. Peters. 1997. Cycle resolved emission spectroscopy for IC engines. SAE paper 972917.
91. Nwagboso, C. 1993. Condition monitoring of CNG engine using optic fibre sensory systems. ISATA 93EN055.
92. Ikeda, Y., A. Nishiyama, N. Kawahara, E. Tomita, S. Arimoto, and A. Takeuchi. 2005. In-spark-plug sensor for analyzing the initial flame and its structure in an SI engine. SAE paper 2005-01-0644.
93. Kawahara, N., E. Tomia, A. Takeuchi, S. Arimoto, Y. Ikeda, and A. Nishiyama. 2005. Measurement of flame propagation characteristics in an SI engine using micro-local chemiluminescence technique. SAE paper 2005-01-0645.
94. Moeser, P. and W. Hentschel. 1996. Development of a time resolved spectroscopic detection system and its application to automobile engines. SAE paper 961199.
95. Pendlebury, M. A. and C. O. Nwagboso. 1996. An optical sensor for determination of combustion parameters in a natural gas fuelled spark ignition engine. SAE paper 960856.
96. Geiser, F., F. Wytrykus, U. Spicher. 1998. Combustion control with the optical fibre fitted production spark plug. SAE paper 980139.
97. Remboski, D. J., S. L. Plee, and J. K. Martin. 1989. An optical sensor for spark-ignition engine combustion analysis and control. SAE paper 890159.
98. Nagase, K., K. Funatsu, Y. Muramatsu, and M. Kawakami. 1985. An investigation of combustion in internal combustion engines by means of optical fibers. SAE paper 851560.
99. Ault, J. R., and P. O. Witze. 1998. Evaluation and optimization of measurements of flame kernal growth and motion using a fiber-optic spark plug probe. SAE paper 981427.
100. Merer, R. M., and J. S. Wallace. 1995. Spark spectroscopy for spark ignition engine diagnostics. SAE paper 9501647.
101. Hultqvist, A., M. Christensen, B. Johansson, A. Franke, M. Richter, and M. Aldén. 1999. A study of the homogeneous charge compression ignition combustion process by chemiluminescence imaging. SAE paper 1999-01-3680.
102. Weimar, H-J., G. Töpfer, U. Spicher. 1999. Optical investigations on a Mitsubishi GDI-engine in the driving mode. SAE paper 1999-01-0504.

103. Beshai, S., A. K. Gupta, S. S. Ayad, and T. A. K. Abdel Gawad. 1990. Chemiluminescence: A diagnostic technique for internal combustion engines. *ASME Applied Fluid Mechanics* 100: 115–18.
104. Nagase, K., K. Funatsu, Y. Muramatsu, and M. Kawakami. 1985. An investigation of combustion in internal combustion engines by means of optical fibers. SAE paper 851560.
105. Augusta, R., D. E. Foster, J. B. Ghandhi, J. Eng, and P. M. Najt. 2006. Chemiluminescence measurements of homogeneous charge compression ignition. SAE paper 2006-01-1520.
106. Yang, J., S. L. Plee, D. J. Remboski, and J. K. Martin. 1990. Comparison between measured radiance and a radiation model in a spark-ignition engine. *Journal of Engineering for Gas Turbines and Power* 112: 331–34.
107. Bach, M., J. Reissing, and U. Spicher. 1996. Temperature measurement and NO determination in SO engines using optical fiber sensors. SAE paper 961922.
108. Boulouchos, K., M. K. Eberle, B. Ineichen, and C. Klukowski. 1989. New insights into the mechanisms of in-cylinder heat transfer in diesel engines. SAE paper 890573.
109. Shoji, H., T. Shimizu, K. Yoshida, and A. Saima. 1995. Spectroscopic measurement of radical behavior under knocking operation. SAE paper 952407.
110. Ziejewski, M., H. J. Goettler, and D. G. Dimitriu. 1991. Development of an infrared method for ignition delay measurements. SAE paper 910847.
111. Shimoda, K., H. Koide, F. Kobayashi, M. Nagase, S. Ikeda, M. Takata, and J. Nakano. 1986. Development of new electronic control system for a diesel engine. SAE paper 860597.
112. Wendeker, M. 1998. Experimental results of the investigation of the mixture preparation in spark ignition engine. SAE paper 982525.
113. Heywood, J. B. 1998. *Internal combustion engine fundamentals.* New York: McGraw-Hill. Pages 572ff.
114. Calcote, H. F. 1957. Mechanisms for the formation of ions in flames. *Combustion and flame* 1: 385–403.
115. Lawton, J. and F. J. Weinberg. 1969. *Electrical aspects of combustion.* Oxford: Clarendon Press.
116. Shimasaki, Y., M. Kanehiro, S. Baba, S. Maruyama, T. Hisaki, and S. Miyata. 1993. Spark plug voltage analysis for monitoring combustion in an internal combustion engine. SAE paper 930461.
117. Mortara, W. and C. Canta. 1983. Engine stability sensor. SAE paper 830428.
118. Witze, P. O. 1989. Cycle-resolved multipoint ionization probe measurements in a spark ignition engine. SAE paper 892099.
119. Clements, R. M., and P. R. Smy. 1976. The variation of ionization with air/fuel ratio for a spark ignition engine. *Journal of Applied Physics* 47: 505–9.
120. Auzins, J., H. Johansson and J. Nytomt. 1995. Ion-gap sense in misfire detection, knock and engine control. SAE paper 950004.
121. Arrigoni, V., F. Calvi, G. M. Cornetti, and U. Pozzi. 1973. Turbulent flame structure as determined by pressure: Development and ionization intensity. SAE paper 730088.
122. Miyata, S., Y. Ito, and Y. Shimasaki. 1993. Flame ion density measurement using spark plug voltage analysis. SAE paper 930462.
123. Griffiths, J. F., and J. A. Barnard. 1995. *Flame and combustion.* 3rd ed. Glasgow, UK: Blackie.
124. Daniels, C. F., and B. M. Scilzo. 1996. The effects of electrode design on mixture ignitability. SAE paper 960606.
125. Douaud, A., G. de Soete, and C. Henault. 1983. Experimental analysis of the initiation and development of part-load combustions in spark ignition engines. SAE paper 830338.
126. Förster, J., A. Günther, M. Ketterer, and K-J. Wald. 1999. Ion current sensing for spark ignition engines. SAE paper 1999-01-0204.
127. Lee, A., and J. S. Pyko. 1995. Engine misfire detection by ionization current monitoring. SAE paper 950003.
128. Dale, J. D., M. D. Checkel, and P. R. Smy. 1997. Application of high energy ignition systems to engines. *Progress in Energy and Combustion Science* 23: 379–98.
129. Reinmann, R., A. Saitzkoff, and F. Mauss. 1997. Local air-fuel ratio measurements using the spark plug as an ionization sensor. SAE paper 970856.
130. Pashley, N., R. Stone, and G. Roberts. 2000. Ignition system measurement techniques and correlations for breakdown and arc voltages and currents. SAE paper 2000-01-0245.

131. Collings, N., S. Dinsdale, and D. Eade. 1986. Knock detection by means of the spark plug. SAE paper 860635.
132. Nicholson, D. E., and P. O. Witze. 1993. Flame location measurements in a production engine using ionization probes embodied in a printed-circuit-board head gasket. SAE paper 930390.
133. VanDyne, E. A., C. L. Burckmyer, A. M. Wahl, and A. E. Funaioli. 2000. Misfire detection from ionization: feedback utilizing the SmartFire plasma ignition technology. SAE paper 2000-01-1377.
134. Brehob, D. D. 1989. An exhaust ionization sensor for detection of late combustion with EGR. SAE paper 892084.
135. Vressner, A., P. Strandh, A. Hultqvist, P. Tunestål, and B. Johansson. 2004. Multiple point ion current: Diagnostics in an HCCI engine. SAE paper 2004-01-0934.
136. Strandh, P., M. Christenen, J. Bengtsson, R. Johansson, A. Vressner, P. Tunestål, and B. Johansson. 2003. Ion current sensing for HCCI combustion feedback. SAE paper 2003-01-3216.
137. Glavmo, M., P. Spadafora, and R. Bosch. 1999. Closed loop start of combustion control utilizing ionization sensing in a diesel engine. SAE paper 1999-01-0549.
138. Kubach, H., A. Velji, U. Spicher, and W. Fischer. 2004. Ion current measurement in diesel engines. SAE paper 2004-01-2922.
139. Pfeffer, T., P. Bühler, D. E. Meier, and Z. Hamdani. 2002. Influence of intake tumble ratio on general combustion performance, flame speed and propagation at a Formula One type high-speed research engine. SAE paper 2002-01-0244.
140. Russ, S., G. Peet, and W. Stockhausen. 1997. Measurements of the effect of in-cylinder motion on flame development and cycle-to-cycle variations using an ionization probe head gasket. SAE paper 970507.
141. Saitzkoff, A., R. Reinmann, F. Mauss, and M. Glavmo. 1997. In-cylinder pressure measurements using the spark plug as an ionization sensor. SAE paper 97085.
142. Docquier, N., and S. Candel. 2002. Combustion control and sensors: A review. *Progress in Energy and Combustion Science* 28: 107–50.
143. Saitzkoff, A., R. Reinmann, T. Berglind, and M. Glavmo. 1996. An ionization equilibrium analysis of the spark plug as an ionization sensor. SAE paper 96033.
144. Quader, A. A., and C. J. Dasch. 1992. Spark plug fouling: A quick engine test. SAE paper 920006.
145. Collings, N., S. Dinsdale, and T. Hands. 1991. Plug fouling investigations on a running engine: An application of a novel multi-purpose diagnostic system based on the spark plug. SAE paper 912318.
146. Kittelson, D. B., and N. Collings. 1987. Origin of the response of electrostatic particle probes. SAE paper no. 870476.
147. Flagan, R. C., and J. H. Seinfeld. 1988. *Fundamentals of air pollution engineering,* 384. London: Prentice Hall.
148. Jiewertz, S., L-O. Ottosson, and J. Bengtsson. 1997. Practical combustion analysis by ionization. *5th CEC International Symposium on the Performance Evaluation of Automotive Fuels and Lubricants,* CEC, Göteborg, Sweden. In Tribotest Journal, Pub Wiley (New York), 5,4.
149. An, F., G. Rizzoni, and D. Upadhyay. 1997. Combustion diagnostics in methane-fueled SI engines using the spark plug as an ionization probe. SAE paper 970033.
150. Peron, L., A. Charlet, P. Higelin, B. Moreau, and J. F. Burq. 2000. Limitations of ionization current sensors and comparison with cylinder pressure sensors. SAE paper 2000-01-2830.
151. Daniels, C. F. 1998. The comparison of mass fraction burned obtained from the cylinder pressure signal and spark plug ion signal. SAE paper 980140.
152. Franke, A., P. Einewall, B. Johansson, N. Wickström, R. Reinmann, and A. Larsson. 2003. The effect of in-cylinder gas flow in the interpretation of the ionization signal. SAE paper 2003-01-1120.
153. Daniels, C. F., G. G. Zhu, and J. Winkelman. 2003. Inaudible knock and partial burn detection using in-cylinder ionization signal. SAE paper 2003-01-3149.
154. Williams, D., P. J. Shayler, and N. Collings. 1988. Exhaust gas ionization sensor for spark ignition engines. *Institution of Mechanical Engineers* C59/88.
155. Lefebvre, C., G. Banet, and A. Ecomard. 1998. Closed loop control of spark-ignition engines: Application to the fuel system. FISITA 885068.

156. Strandh, P., J. Bengtsson, R. Johansson, P. Tunestål, and B. Johansson. 2004. Cycle-to-cycle control of a duel-fuel HCCI engine. SAE paper 2004-01-0941.
157. Schneider, D., and M-C. Lai. 2000. An investigation of the impact of cycle-to-cycle variations on the ionic current signal in SI engines. SAE paper 2000-01-1943.
158. Yoshiyama, S., E. Tomita, K. Matsumoto, and K. Matsuki. 2003. Combustion diagnostics of a spark ignition engine by using gasket ion sensor. SAE paper 2003-01-1801.
159. Anderson, R. I.. 1986. In-cylinder measurement of combustion characteristics using ionization sensors. SAE paper 860485.
160. Rado, W. G., and W. J. Johnson. 1975. Significance of burn types, as measured by using the spark plugs as ionization probes, with respect to the hydrocarbon emission levels in S.I. engines. SAE paper 750354.
161. Schiefer D., Maennel R., Nardoni, W. 2003. Advantages of diesel engine control using in-cylinder pressure information for closed-loop control. SAE paper 2003-01-0364.
162. Eriksson, L., L. Nielsen, and J. Nytomt. 1996. Ignition control by ionization current interpretation. SAE paper 960045.
163. Yoshima, S., E. Tomita, K. Matsumoto, and K. Matsuki. 2003. Combustion diagnostics of a spark ignition engine by using gasket ion sensor. SAE paper 2003-01-1801.
164. Pestana, G. W. 1989. Engine control methods using combustion pressure feedback. SAE paper 890758.
165. Ishida M., H. Ueki, Y. Yoshimura, and N. Matsumura. 1990. Studies on combustion and exhaust emissions in a high speed DI diesel engine. SAE paper 901614.
166. Kawahara, N., E. Tomita, and H. Kagajyo. 2003. Homogeneous charge compression ignition combustion with dimethyl ether—spectrum analysis of chemiluminescence. SAE paper 2003-01-1828.
167. Mayer, R., J. T. Kubesh, S. M. Shahed, and J. K. Davies. 1993. Simultaneous application of optical spark plug probe and head gasket ionization probe to a production engine. SAE paper 2003-01-1117.
168. Huang, Y. and D. Mehta. 2005. Investigation of an in-cylinder ion sensing assisted HCCI control strategy. SAE paper 2005-01-0068.
169. Sohma, K., T. Yukitake, S. Azuhata, and Y. Takaku. 1991. Application of rapid optical measurement to detect the fluctuations of the air-fuel ratio and temperature of a spark ignition engine. SAE paper 910499 (1991)
170. Ikeda, Y., H. Nishihara, and T. Nakajima. 2001b. Measurement of flame front structure and its thickness by planar and local chemiluminescence of OH*, CH* and C2*. SAE paper 2001-01-0920.
171. Kerstein, A. R. and P. O. Witze. 1990. Flame-kernel model for analysis of fiber-optic instrumented spark plug data. SAE paper 900022.
172. Spicher, U. and H. Backer. 1990. Correlation of flame propagation and in-cylinder pressure in a spark ignited engine. SAE paper 902126.
173. Winklhofer, E., C. Beidl, H. Philipp, and W. F. Piock. 2001. Micro-optic sensor techniques for flame diagnostics. Paper presented at the JSAE Spring Conference, Yokohama, Japan.
174. Byttner, S., T. Rögnvaldsson, and N. Wickström. 2001. Estimation of combustion variability using in-cylinder ionization measurements. SAE paper 2001-01-3485.
175. Müller, R., R. Hart, G. Krötz, M. Eickhoff, A. Truscott, A. Noble, C. Cavalloni, and M. Gnielka. 2000. Combustion pressure based engine management. SAE paper 2000-01-0928.
176. Ohashi, Y., W. Fukui, and A. Ueda. 1997. Application of vehicle equipped with ionic current detection system for the engine management system. SAE paper 970032.
177. Ohashi, Y., W. Fukui, F. Tanabe, and A. Ueda. 1998. The application of ionic current detection system for the combustion limit control. SAE paper 980171.
178. Panousakis, D., A. Gazis, J. Patterson, and R. Chen. 2006. Analysis of SI combustion diagnostics methods using ion-current sensing techniques. SAE paper 2006-01-1345.
179. Checkel, M. D. and J. D. Dale. 1989. Pressure trace knock measurement in a current S.I. production engine. SAE paper 890243.
180. Chae, J. O., A. A. Martychenko, S. W. Kim, S. C. Chung, G. M. Vasilev, and A. V. Yurkevich. 1994. An application study of the system for knocking-effect control on real car. Paper presented at the SAE Spring Conference, Seoul, Korea, 41–46.

181. Millo, F. and C. V. Ferraro. 1998. Knock in S.I. engines: A comparison between different techniques for detection and control. SAE paper 982477.
182. Sawamoto, K., Y. Kawamura, T. Kita, and K. Matsushita. 1987. Individual knock control by detecting cylinder pressure. SAE paper 871911.
183. May, M. G. 1984. Flame arrival sensing response double closed loop engine management. SAE paper 840441.
184. Ohashi, Y., M. Koiwa, K. Okamura, and A. Ueda, 1999. The application of ionic current detection system for the combustion control condition. SAE paper 1999-01-0550.
185. Harda, J., H. Watanabe, K. Nakanishi, and T. Kawai. 1993. A new generation of Toyota lean burn engine: Worldwide emission standards and how to meet them. *Institution of Mechanical Engineers*, pp. 31–36.
186. de Bie, T., M. Ericsson, and P. Rask. 2000. A novel start algorithm for cng engines using ion sense technology. SAE paper 2000-01-2800.
187. Galindo, J., V. Bermúdez, J. R. Serrano, and J. López. 2001. Cycle to cycle diesel combustion during engine transient operation. SAE paper 2001-01-3262.
188. Müller, R. and R. Isermann. 2001. Control of mixture composition using cylinder pressure sensors. SAE paper 2001-01-3382.
189. Patrick, R. S. and J. D. Powell. 1990. A Technique for the real-time estimation of air-fuel ratio using molecular weight ratios. SAE paper 900260.
190. Yoon, P., S. Park, M. Sunwoo, I. Ohm, and K. J. Yoon. 2000. Closed-loop control of spark advance and air-fuel ratio in SI engines using cylinder pressure. SAE paper 2000-01-0933.
191. Ikeda, Y., M. Keneko, and T. Nakajima. 2001a. Local A/F measurement by chemiluminescence OH*, CH* and C2* in SI engine. SAE paper 2001-01-0919.
192. Nogi, T., J. Yamahuchi, M. Ohsuga, and N. Kurihara. 1992. Effects of mixture formation technology on gasoline engine performance. SAE paper 922092.
193. Eriksson, L., L. Nielsen, and M. Glavenius. 1997. Closed loop ignition control by ionization current interpretation. SAE paper 970854.
194. Shamekhi, A. H. and A. Ghaffari. 2005. Fuzzy control of spark advance by ion current sensing. SAE paper 2005-01-0079.

CHAPTER 6

AUTOMOTIVE TORQUE SENSORS

John Turner

1. INTRODUCTION

Torque is frequently a cause of confusion, even among experienced engineers. Statements such as "torque is the amount of twist in the shaft" or "torque is a measure of the windup" are often heard. It is sensible therefore to begin with a formal definition of torque, before moving on to consider methods by which it can be measured.

Torque can be defined as a measure of the tendency of a force to rotate the body on which it acts about an axis. Everyday experience tells us that the "rotating effectiveness" of a force increases with its perpendicular distance from the pivot. For example, when opening a door it is sensible to push or pull as far as possible from the hinges, and we attempt to keep the direction of the pull or push perpendicular to the door.

The magnitude of the torque acting in a plane perpendicular to an axis is obtained by multiplying the force (or the component of the force in a plane perpendicular to the axis of rotation), by the perpendicular distance from the axis to the line of action of the force, as shown in figure 6.1. The SI units of torque are Newton-meters (nm).

The discussion above applies to both static problems (where, for example, a structural member is subjected to a moment) and to the more important case (for engineers) of rotating shafts in torsion. For the automotive engineer, torque in a rotating shaft is important, since it is the limiting factor in determining how much power the shaft can transmit. A shaft rotating with angular velocity ω and carrying power P will undergo a torque T, where

$$T = \frac{P}{\omega}. \qquad (6.1)$$

A shaft of length l, polar moment of inertia J, and modulus of rigidity G, subjected to a torque T, will experience an angle of twist θ given by

$$\theta = \frac{Tl}{JG}. \qquad (6.2)$$

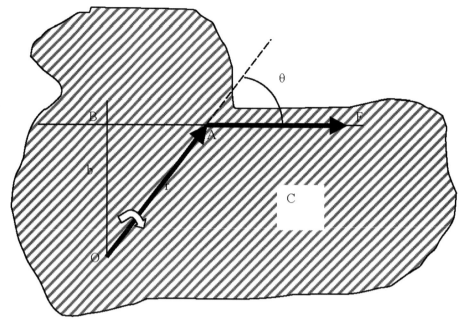

Figure 6.1. Definition of torque: Force F acting at point A on body C tends to rotate C about O. Torque $T = Fb = Fr\sin\theta$. In vector notation $\underline{T} = \underline{r} \times \underline{F}$.

The maximum shear stress τ occurs at the surface of the shaft and is

$$\tau = \frac{Tr}{J},\quad (6.3)$$

where r is the shaft radius as shown in figure 6.2. For a solid circular shaft, the polar moment of inertia is $J = \pi r^4/2$, so by substitution we have

$$\theta = \frac{2Tl}{\pi_r^4 G} \quad \text{and} \quad \tau = \frac{2T}{\pi_r^3}. \quad (6.4)$$

Equations for the strains produced by torsion in shafts of other-than-circular cross sections are readily derived [1].

The most important application of torque measurement in automotive engineering is in the assessment of engine power. Equation 6.1 shows how power can readily be calculated from measurements of rotational speed and torque. Engine speed measurement is relatively easy—often a white paint mark on the flywheel and a simple optical reflectance sensor will suffice. Torque measurement is more difficult to arrange, and for this reason it is still in the main the preserve of research and development laboratories. However, there are signs that this situation may be about to change, and torque sensors may soon become commonplace on production vehicles.

The designers of powertrain management systems would find a direct measurement of torque very useful. Strain-gauged torque sensors (see next section) are available for use in development work, but are far too expensive (and unreliable in the long term) for use in production vehicles.

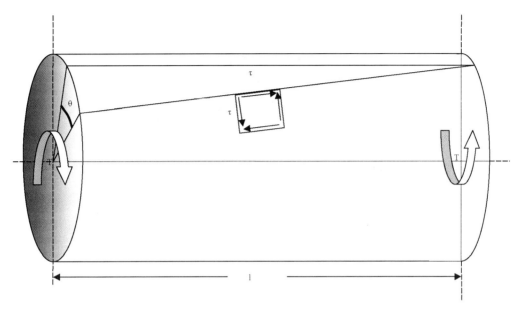

Figure 6.2. Circular shaft (radius *r*, length *l*) under torque *T* produces shear stress τ.

The utility of a reliable torque measurement device is apparent from consideration of the torque-speed characteristics of an engine and gearbox in combination. The gearbox is fitted between the engine and the road wheels of a vehicle, and serves purely as an impedance matching device. With a conventional manual gearbox the driver acts within the feedback loop, sensing speed and load, and adjusting the transmission ratio to what appears to be the optimum setting. The main sensory input used to achieve optimization is engine speed, perceived in the form of noise. Unfortunately, the pitch and noise level of an engine are poorly correlated with its power output or efficiency, so although gear changing controlled by acoustic stimuli may optimize the subjective acceleration and drivability, it gives poor economy and performance [2]. Figure 6.3 shows the torque-speed curves for a typical engine and is taken from reference 2. From these curves it is apparent that the optimum economy is obtained by keeping the engine at as low a speed as possible during acceleration (line A–B on figure 6.3), while using the gears to increase vehicle speed. The engine speed is only raised to increase the vehicle speed when the final gear ratio is reached at point B. Subsequently, the powertrain is controlled for optimum fuel economy and operates along line B–C. Good acceleration performance of course requires a somewhat different strategy.

2. MECHANICAL METHODS OF TORQUE MEASUREMENT

One of the earliest (and still very useful) methods of measuring the torque produced by an engine uses a device known as an absorption dynamometer, in which all the power produced by the engine is absorbed by friction in a brake. This is the origin of the phrase "brake horsepower," although shaft power is a less misleading term. The engine is fitted with a rope or belt brake wrapped around the flywheel, which is often water cooled. The rope passes once around the flywheel and is attached to a

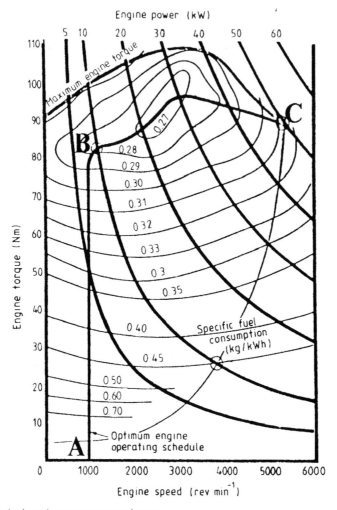

Figure 6.3. Typical engine torque-speed curves

mass M at the bottom as shown in figure 6.4. The other end of the rope is connected to a spring balance, which measures the tension in the rope F. The force in the lower end of the rope arises from the weight, and is Mg. If the spring balance reading is F, the difference in tension between the ends of the rope is $(Mg - F)$. If the radius of the flywheel is r, the torque will be

$$Torque = (Mg - F)r \qquad (6.5)$$

and the power

$$Power = 2\pi N(Mg - F)r \qquad (6.6)$$

(where N is the number of flywheel revolutions per second).

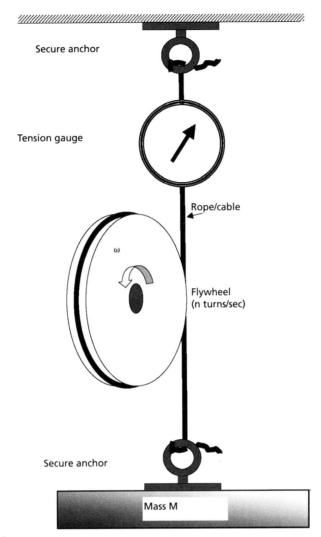

Figure 6.4. Brake dynamometer

The danger inherent in this arrangement is that the brake may jam, throwing the weight over the top of the flywheel. To avoid this alarming possibility a strong safety rope or chain is always used to prevent the weight being lifted more than a few centimeters.

A more sophisticated brake dynamometer frequently used for torque measurement in automotive powertrain research is the hydraulic type, originally invented by W. Froude. In a Froude dynamometer the energy from a rotating shaft is transferred to water contained within a cylindrical housing. Shear forces within the water then attempt to rotate the housing. The torque required to restrain the device is measured by (usually) a spring balance. The advantage of a Froude dynamometer is that unlike a rope brake device there is no possibility of it "snatching." However, Froude dynamometers are more expensive than rope brakes. It is interesting to note in passing that the automotive fluid flywheel was almost certainly developed from the Froude dynamometer.

An entirely mechanical method of torque measurement is based on a measurement of the force required to restrain a gearbox. As far as the authors are aware, it has not been applied in automotive engineering, although there seems to be no reason why it should not be successful. Any gearbox that changes the rotational speed of a shaft will change the torque in inverse proportion (assuming friction can be neglected). The ratio of the input torque T_{in} to output torque T_{out} is equal to the reciprocal of the speed ratio, and the *difference* between the input and output torques is the torque needed to restrain the gearbox. Thus

$$\frac{T_{in}}{T_{out}} = \frac{\omega_{out}}{\omega_{in}} \qquad (6.7)$$

and

$$(T_{in} - T_{out}) = T_{restraining} \qquad (6.8)$$

By measuring the input and output speeds, and the restraining torque, T_{in} and T_{out} can be calculated. This approach has been found to be very useful in university laboratories where costs are of overriding importance, since it allows a torque measuring system to be improvised cheaply using a scrap back axle from a rear-wheel-drive vehicle. The engine is connected to one wheel shaft and the load to the other. The propeller shaft coupling is locked to the differential housing. The torque required to restrain the housing is measured with a spring balance or electronic force transducer, and is twice the torque being transmitted through the system since the input and output shafts revolve in opposite directions.

3. STRAIN-GAUGE TORQUE TRANSDUCERS

Strain-gauge torque transducers are created by applying strain gauges to a shaft to measure the shear strain caused by torsion, as shown in figure 6.2 and discussed in section 1. They are very widely used in research laboratories and are probably the most common form of torque sensor. Their major disadvantage is that they require additional equipment to transmit power to the rotating shaft, and to retrieve data from it. This can take the form of a set of slip rings, rotary transformers, or battery-powered radio telemetry equipment. Regardless of which of these is chosen, the need for some form of power and/or data transmission system, and the consequential costs incurred, probably rules out strain gauge-based torque sensors for use on volume-produced vehicles. In addition slip rings (and to some extent rotary transformers) can be unreliable when operated in a dirty environment and may be prone to radio-frequency interference (RFI).

The shear stresses illustrated in figure 6.2 cause strains to appear at 45° to the longitudinal axis of the shaft. The conventional arrangement of strain gauges for torque measurement is shown in figure 6.5. The gauges must be placed precisely at 45° to the shaft axis; otherwise the arrangement is sensitive to bending and axial stresses in addition to those caused by torsion. Accurate gauge placement is facilitated by the availability of special "rosettes" in which two gauges are precisely positioned on a common backing. The use of four active strain gauges in a bridge arrangement gives complete thermal compensation [3]. Figure 6.5 shows an arrangement of strain gauges on a solid circular shaft. The same gauge positioning can be used on a hollow circular shaft, such as the propeller shaft on a rear-wheel-drive vehicle. When torque is applied to a thin-walled cylinder, such as a rear-wheel-drive shaft,

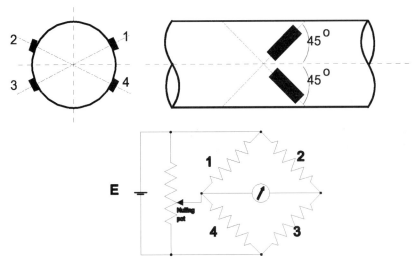

Figure 6.5. Strain gauges for shaft torque measurement

the shear stress is assumed to be constant throughout the wall [4]. In such cases it is often convenient to place the strain gauges on the inner surface of the driveshaft where they are afforded a degree of mechanical protection.

Shafts of other-than-circular cross sections are sometimes used for torque measurement as shown in figure 6.6. For measuring low levels of torque, the cruciform or hollow cruciform configuration is sometimes used. A solid square shaft is suitable for larger torque values and has a number of advantages over the circular shaft of figure 6.5. The strain gauges are more easily aligned and attached to a flat surface, and since the corners of a square section in torsion are stress free [5], they provide a good location for the solder joints between leads and strain gauges. These joints are often a source of unreliability due to fatigue failure if they are located in a high-stress region. Finally, a square shaft is much stiffer in bending than a circular one of equivalent torsional stiffness, so the effects of bending (which will appear if the gauges are misaligned) are reduced.

4. TORSION BARS

Torque in a shaft leads to elastic deflection. The resulting strain can be measured at a point as described in the preceding section, or alternatively the gross relative motion between the ends of the shaft may be used to indicate the torque. Just as in the case of strain-gauge systems, a major difficulty is the necessity of being able to measure the deflection while the shaft is rotating. However, there are advantages in using shaft deflection. First, the need for precise location and orientation of the strain sensors is avoided. Second, since the effect of an applied torque is integrated along the length of the shaft, the influence of any local variation in material properties or shaft geometry is reduced. Third, the (relatively) larger displacements available when movements of the two ends of a shaft are compared make it possible to design a variety of noncontact torque measurement systems, which avoid the need for slip rings.

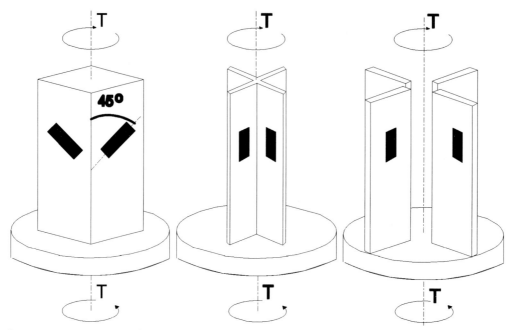

Figure 6.6. Torque sensor designs

Figure 6.7 shows a typical torsion-bar torque meter using an optical method for deflection measurement. The relative angular displacement between the ends of the torque-transmitting member is read from the position of the pointer on disk two relative to the calibrated scale fixed to disk one. The "persistence" of human vision and the stroboscopic effect of intermittent viewing make it possible to operate this system from about 600 rpm (10 Hz) upwards.

A torsion-bar system using capacitive torque sensing has been demonstrated for automotive use [6]. An automotive driveshaft fitted with a concentric sleeve of dielectric material is shown in figure 6.8. The sleeve is fixed to the shaft at one end and rests on a rubbing bearing at the other end. When torque is applied to the shaft, it causes relative motion between the surface of the shaft and the free end of the concentric tube. This motion is used to vary the capacitance between two opposing patterns of conducting strips, one of which is applied to the shaft and one to the tube. The capacitive torque sensor is connected to an inductor coil wound around the shaft. The resulting passive circuit has a resonance frequency that depends on the applied torque. The passive resonant circuit rotates with the driveshaft and is excited from an adjacent stationary location by inductive coupling using a second inductor coil driven by an oscillator as shown in figure 6.9. The problem of torque measurement then becomes one of measuring the frequency at which resonance occurs. When the oscillator frequency is the same as that at which resonance occurs in the passive circuit, an increased current is drawn. If the frequency at which this occurs is measured, it can be used to indicate the torque. The advantage of this arrangement for automotive applications is that no physical connection between the rotating shaft and the vehicle body is required.

The sensors in the prototype device were manufactured by thick-film printing and were found to be reasonably robust. In the prototype signal conditioning system, the torque sensor was inductively

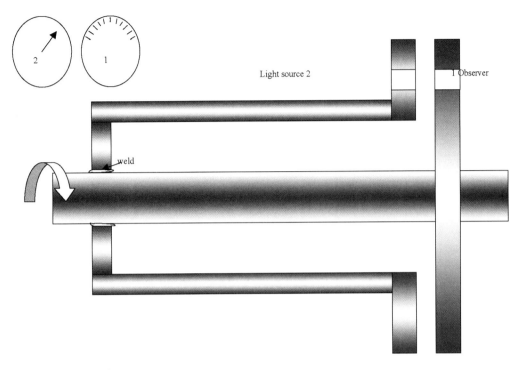

Figure 6.7. Torsion-bar torque meter

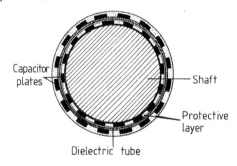

a) Cross section through shaft and sensor

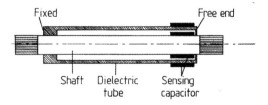

b) Longitudinal section

Figure 6.8. Construction of capacitive driveshaft torque sensor

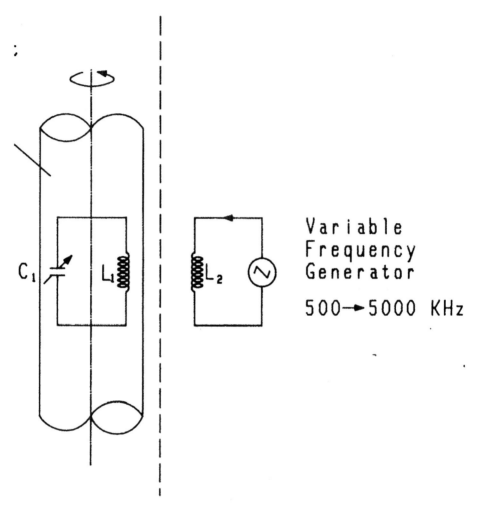

Figure 6.9. Rotating measurement circuit excited by inductive coupling

coupled into a capacitive bridge as shown in figure 6.10. With this arrangement the output was found to vary as shown in figure 6.11. It can be seen that it is reasonably linear and is probably adequate for automotive applications, such as engine management or automatic gearbox control.

The prototype used a relatively short driveshaft from a front-wheel-drive vehicle (a Ford Escort). There is no reason why the same technique could not be applied to rear-wheel-drive vehicles, and in fact it may be easier, since the driveshafts used tend to be longer and will have a larger "windup."

An optical torsion-bar sensor intended for use as part of an electric power-assisted steering (EPAS) system has been proposed by the Lucas Advanced Engineering Centre [7]. The optical sensor uses a pair of slotted disks positioned at the ends of a torsion bar as shown in figure 6.12. Light from a light-emitting diode (LED) passes through the slots in the disks and is received by a photodetector.

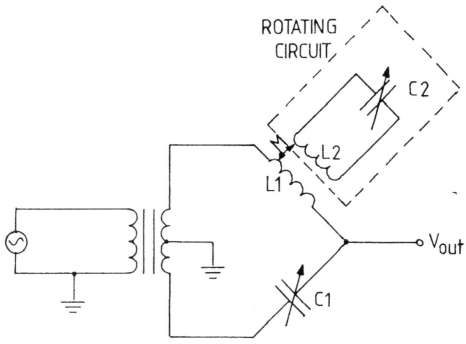

Figure 6.10. Bridge circuit for telemetry

Torque variations cause the amount of overlap between the disks to vary and hence the output from the photodetector. However, the Lucas system exhibits a number of refinements, which are intended to make it more suitable for automotive use. The most important of these is the use of a ratiometric technique to cancel out the effect of any variation in the source illumination intensity. The slotted disks are illuminated by a common LED as shown in figure 6.12. The amount of light emitted by the diode will vary if the supply voltage changes. Even if a well-regulated supply is available, the light output from an LED reduces by up to 40% as the device ages. The ratiometric effect is achieved by arranging for each slotted disk to carry two tracks of slots positioned so that as torque is applied in one direction the light intensity transmitted through the outer track (A) increases, while that passing through the inner track (B) decreases. The light passing through each track is measured by a pair of photodiodes as shown in figure 6.12. The torque is calculated by measuring the outputs from photodiodes A and B and then evaluating the expression

$$Torque \propto \frac{A-B}{A+B}. \qquad (6.9)$$

The magnitude of the result gives the torque, and the sign, the direction (i.e., clockwise or counterclockwise) in which it is applied. Provided both channels are affected equally, this technique ensures that the torque measurement is independent of the source intensity. Furthermore, for a given source intensity the quantity (A + B) should be a constant, which is independent of torque, and this value can be used to check that the sensor is operating correctly. If (A + B) moves outside preset limits, an

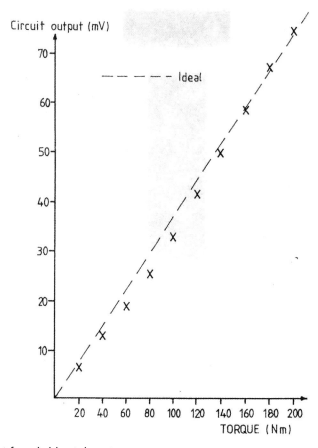

Figure 6.11. Output from bridge telemetry

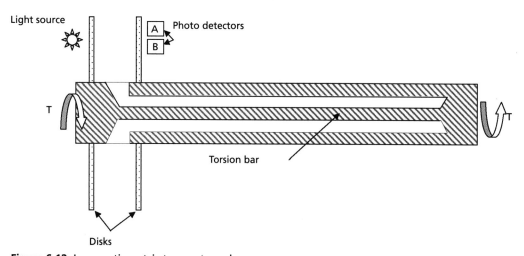

Figure 6.12. Lucas ratiometric torque transducer

appropriate warning may be given. A self-test facility of this kind is obviously essential in a safety-critical system such as vehicle steering.

The main problem with the Lucas system appears to be that the geometry of the photodetectors, and their location with respect to the slots on the disks, is critical if ripple in the sensor output is to be prevented as the disks rotate. Variations in the output can be avoided only if the sensitive area of the light detectors corresponds exactly to an even multiple of the slot area. Reference 7 proposes the use of masks to give the correct detection area and to collimate the light source. The ripple amplitude after these improvements is reported to be better than 10% of the full-scale measurement range. Although this level of accuracy would not be acceptable for a laboratory torque sensor, it is probably adequate for power steering applications.

Work on measuring the twist or "windup" along the crankshaft of an engine using slotted disks at each end has also been reported [8]. However, the very high levels of torque variation that result from multicylinder engine operation are alleged to make it difficult to obtain accurate results.

5. NONCONTACT MAGNETIC METHODS

A number of torque sensors utilizing the magnetostrictive effect have been reported. A good example of this approach is a device described by Spectrol Electronics [9] and is shown in figure 6.13. Magnetostriction is an effect that occurs in ferromagnetic materials such as steel, where the magnetic permeability is affected by stress. Equation 6.3 shows that the stress in a shaft is proportional to the applied torque, and it follows that torque must change the permeability of the shaft if it is made of a magnetic material such as steel. The effect is small but can be measured by an arrangement such as that shown in figure 6.13. The torque sensor consists of five coils arranged as shown, wound onto a common five-armed core. The center coil can be thought of as the primary winding of a transformer, and the four

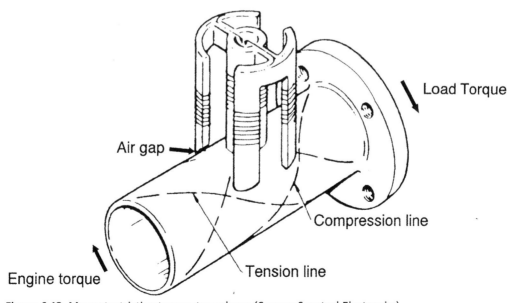

Figure 6.13. Magnetostrictive torque transducer (*Source:* Spectrol Electronics)

circumferentially positioned coils act as secondaries. Magnetic coupling between the primary and the secondaries is provided by the steel shaft, which is positioned close to the sensor as shown. The primary coil is excited by an AC current and produces an oscillating magnetic field within the shaft. The four secondary coils are connected together in a Wheatstone bridge arrangement and are positioned so that they lie over the lines of principal stress, which follow a helical path at 45° for a cylinder in torsion. When the shaft is not under torsion, equal currents are induced in the four secondaries, and the bridge out-of-balance voltage is zero. When torque is applied to the shaft, the permeability in the tension and compression directions will change by equal but opposite amounts, and the amplitude of the resulting bridge output voltage is proportional to the applied torque.

There are four main problems with this type of torque sensor:

- Inhomogeneity of the shaft material
- Sensitivity to changes in the sensor/shaft gap
- Thermal effects
- Variations in the sensor output due to changes in the shaft rotation speed

The first of these effects is the most serious. The permeability of the material from which the shaft is made can vary by up to 50% around the circumference of the shaft. For a constant torque the output signal from the sensor can "ripple" as a result at a frequency equivalent to the rotation rate. This characteristic makes it very difficult to measure instantaneous torque levels around a rotating shaft. However, the use of smoothing circuits allows the device to be used for measuring the average torque in the shaft by integrating over several revolutions.

Variable-permittivity torque sensors of this type are probably not suitable for use as part of a high-speed real-time engine management system where instantaneous torque measurement is required, but they may be useful for less demanding tasks such as automatic transmission control.

5.1. SURFACE ACOUSTIC WAVE (SAW) TORQUE TRANSDUCERS

SAW devices are based on a theory propounded by Lord Rayleigh in 1885 [10], which showed that waves (known as surface waves or Rayleigh waves) could propagate along the surface of an isotropic elastic medium. Surface acoustic waves can be excited and detected using piezoelectric transducers etched with a pattern of interdigitated electrodes as shown in figure 6.14. The frequency at which the SAW device operates is determined by the electrode geometry. For torque measuring, two SAW devices are attached to the shaft undergoing torsion. The shear strain resulting from torque changes the geometry of the electrodes, and hence the operating frequency of the device. At a frequency of 500 MHz, 1000 $\mu\varepsilon$ (microstrains) will alter the SAW frequency by 500 kHz [11]. The two transducers are positioned on the shaft at 45°. Each transducer forms part of the feedback loop in an oscillator, such that the output frequency is a function of the SAW geometry. The two SAW transducers are used in a half-bridge configuration, one undergoing tension and the other compression. The resulting two frequencies are added or subtracted: the difference in frequency gives a measure of torque, and the sum can be used to estimate temperature.

The SAW devices can be driven without the need for any electrical connection if capacitive or inductive pickups are used. This feature makes SAW-based systems particularly attractive for automotive torque measurement, since (as discussed at the start of this section) telemetry based on slip rings, or rotary transformers can be a source of unreliability and is often too expensive for use in mass-produced vehicles.

6. SUMMARY

In conclusion, it appears that although automotive powertrain engineers would find a low-cost, reliable torque sensor very useful for engine and transmission control, one has yet to be produced that meets enough of the requirements to gain widespread acceptance. While strain-gauge systems using sliding contacts or telemetry are adequate for development work, they are prohibitively expensive for use on production vehicles. The torsion-bar systems described in section 4 have a suitable performance, can be low-cost, and avoid the need for telemetry. However, the optical versions are liable to dirt problems, which may prohibit their use for powertrain applications. It is probably feasible, however, to use optical systems in "clean" applications such as steering torque measurement. The capacitive torque sensor described in reference 6 does not need to be kept as clean as an optical system, but is rather labor-intensive to construct and requires careful setting up. Further development may reduce these disadvantages.

The magnetostrictive system described in section 5 also avoids the need for telemetry or an electrical connection to the rotating shaft, but suffers from the drawbacks discussed earlier. In addition an experimental version constructed for evaluation purposes was found to be very temperature sensitive.

To sum up, it appears that torque sensing is a real need in automotive engineering, but to date no practical low-cost systems have been developed which are suitable for widespread use.

REFERENCES

1. Roark, R. J., and W. C. Young. 1976. *Formulas for stress and strain*. 5th ed. New York: McGraw-Hill.
2. Westbrook, M. H. 1988. Automotive transducers: An overview. *Proceedings of IEE Part D* 135(5): 339–47.
3. Turner, J. D., and M. Hill. 1999. *Instrumentation for engineers and scientists*. Oxford: Oxford University Press.
4. Benham, P. P., and R. J. Crawford. 1987. *Mechanics of engineering materials*. London: Longman.

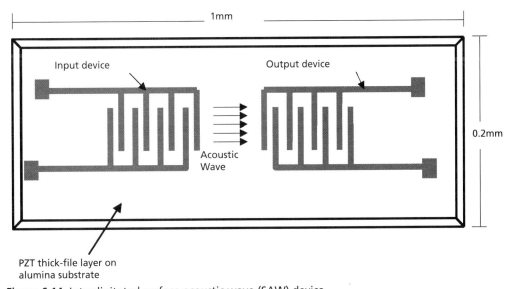

Figure 6.14. Interdigitated surface acoustic wave (SAW) device

5. Timoshenko, S. P., and J. N. Goodier. 1982. *Theory of elasticity*. 3rd ed. New York: McGraw-Hill.
6. Turner, J. D. 1989. The development of a thick-film non-contact shaft torque sensor for automotive applications. *Journal of Physics E: Scientific Instrumentation* 22: 82–88.
7. Hazelden, R. J. 1992. Application of an optical torque sensor to a vehicle power steering system. *Proceedings of IEE Colloquium* C12: 1992/107.
8. Westbrook, M. H. 1985. Sensors for automotive applications. *Journal of Physics E: Scientific Instrumentation* 18: 751–58.
9. Wells, R. F. 1988. Non-contacting sensors for automotive applications. SAE paper 880407.
10. Rayleigh, Lord. 1885. On waves propagated along the plane surface of an elastic solid. *Proceedings of London Mathematical Society* 7: 4–11.
11. Shelley, T. 1993. Acoustics sense torque at low cost. *Eureka*, September, 48–49.

CHAPTER 7

DISPLACEMENT AND POSITION SENSORS

Jonathan Swingler

1. INTRODUCTION

Displacement and position sensors are probably the most widely used type in automotive engineering [1,2]. They are used to monitor a large range of parameters, including the indication of an open door, the level of fluids such as fuel, the spatial position or velocity of suspension system components, and the speed of rotating components.

Probably the first displacement sensor to be fitted to a motor vehicle was the dashboard-mounted gas gauge, which first appeared in the 1920s [3]. This consisted of a voltmeter controlled by the signal generated from a float-and-arm arrangement in the gas tank. The arm moved the wiper of a wire-wound potentiometer to generate a signal [4].

This chapter introduces the technologies used to measure linear and angular position and displacement. Particular sensing technologies are presented in terms of the physical phenomena being utilized by the sensor. A discussion of the particular measurands of interest for automotive applications is presented.

2. LINEAR AND ANGULAR POSITION AND DISPLACEMENT SENSING

Displacement and position sensors can be classified using different methods. These include whether the measurand is linear or angular, whether the output information is absolute or relative, and whether the measurand represents a position, a displacement, or a speed.

2.1. LINEAR AND ANGULAR SENSORS

Two types of mechanical movement are usually monitored for automotive systems: linear and angular movement, as illustrated in figure 7.1.

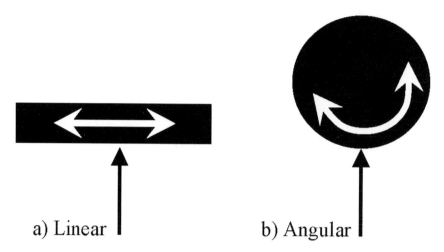

Figure 7.1. Linear and angular movement

The term "linear" can be confusing in the context of sensors as it has two quite different meanings. The first meaning, used in this section, denotes translational rather than angular movement. However, the second meaning of linear describes how the electrical output of a sensor relates to the measurand. An output signal is described as "linear" if the electrical output is linearly related to the measurand. (An alternative might be a logarithmic signal.)

Therefore there are two different definitions for the term "linearity": linearity with respect to the measurand, and output linearity.

2.2. ABSOLUTE AND INCREMENTAL SENSORS

Position information can be presented in one of two ways. Incremental sensors measure a position by counting signal outputs (often in the form of pulses) from the sensor starting at an arbitrary index point. Absolute sensors give an output that is unique to that position. Figure 7.2 illustrates the difference between these two types of sensors.

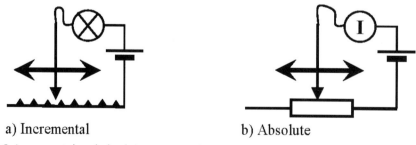

Figure 7.2. Incremental and absolute movement

The incremental sensor shown consists of a displacement contact and a stationary strip of contact points. As the displacement contact moves over the strip of contacts it makes and breaks an electrical circuit. The bulb illustrated in the figure is alternatively lit and extinguished.

The absolute sensor shown consists of a sliding contact on an electrically resistive material (a variable resistor). As the sliding contact moves the resistance of the circuit changes, giving a unique (or absolute) current value for each contact position.

2.3. POSITION, DISPLACEMENT, AND VELOCITY (PDV) SENSORS

In addition to measuring linear/angular or absolute/incremental movement, the sensors described in this chapter can be divided into three general types. The transducer may measure the following:

- Position s (i.e., the location of an item as a numerical value)
- Displacement ds (which gives a change in the location of an item as a numerical value)
- Velocity ds/dt (which indicates the rate of movement)

An example of the s-type measurand in automotive engineering is throttle position as measured by a potentiometric sensor. An example of the ds-type is a change in position, such as a suspension movement, which might be measured by an inductive sensor. An example of the ds/dt (rate) type of measurand is the rotary speed of the engine, as measured by an inductive sensor.

3. SENSOR TECHNOLOGIES

Position, displacement, and velocity sensors convert mechanical measurands to electrical form by various means. The approaches used include potentiometric, inductive, capacitive, and optical methods.

3.1. POTENTIOMETRIC SENSORS

The potentiometric sensor consists of a moveable contact on an electrically resistive component [3]. The earliest versions were made from a resistive wire (such as Nichrome) wound onto an insulating former. The moveable contact, or wiper, makes contact with the wound wire and slides along the wire. Both linear and angular forms can be obtained, with a linear movement normally between 5 mm and 1000 mm, and angular movement between 10° and 21600° (60 turns) [2].

The circuit diagram of a potentiometric sensor made from a resistance R_s with a wiper is shown in figure 7.3. R_s is connected to a power supply with a voltage drop of e_i and a signal conditioning circuit with resistance of R_m giving a voltage output of e_o.

Assuming the potentiometric sensor in figure 7.3 gives a linear output with respect to the displacement of the wiper, the output voltage e_o will be a linear function of displacement x_o when R_m is open circuit ($R_m = \infty$). Therefore equation 7.1 applies

$$\frac{e_o}{e_i} = \frac{x_o}{x_i},$$
(7.1)

where x_i is the full displacement distance on R_s to give the voltage drop of e_i.

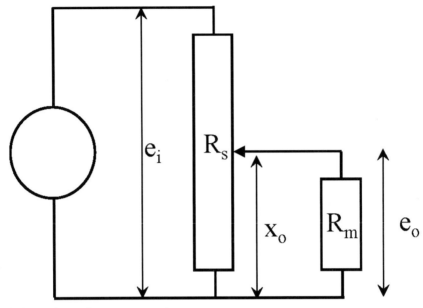

Figure 7.3. Potentiometric measurement

However, when a current is drawn from the sensor by R_m, a signal conditioning circuit, the linearity degrades and equation 7.2 holds.

$$\frac{e_o}{e_i} = \frac{1}{(x_i/x_o) + (R_s/R_m)(1 - x_o/x_i)} \qquad (7.2)$$

Equation 7.2 is plotted in figure 7.4, which illustrates the degradation in signal output linearity as R_m decreases.

To achieve close to linearity R_m should be high compared to the sensor resistance R_s, which should be kept as low as possible. However, this requirement conflicts with the need for high sensitivity of the sensor. For high sensitivity, the supply voltage e_i should be as high as possible since the output signal e_o is directly proportional to e_i. However, a maximum e_i should be found due to the heat-dissipating capability of the sensor. If the maximum heat dissipation is given by H (Watts) the maximum permissible voltage e_i can be found from equation 7.3:

$$e_o(\max) = \sqrt{HR_s} \qquad (7.3)$$

A low value for R_s gives a small e_o and a reduced sensitivity. Thus, the selection of R_s is a compromise between signal output linearity and sensitivity.

The resolution of a potentiometric sensor depends on its construction. If it is a wire-wound and wiper type, the variation in resistance between each winding of the wire as the wiper moves across will have a finite value. This value will be the resolution of that type of sensor. The finer the wire, the higher the resolution achieved. A sensor with 25 turns/mm would have a resolution of ±40 μm [2].

Electrical noise is common with potentiometric sensors due to output voltage fluctuations as the wiper moves across the resistive surface. These voltage fluctuations or contact resistance fluctuations

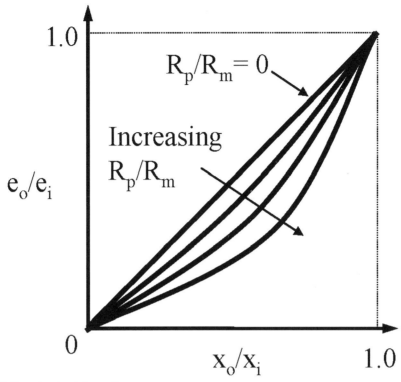

Figure 7.4. Signal output linearity

can be caused by wiper bounce, dirt, or wear at the contact surface. Electrical noise can be minimized by having a multicontact wiper configuration.

A summary of potentiometric sensor characteristics is given in table 7.1.

Table 7.1. Potentiometric sensor summary characteristics

Parameter	Characteristic
Sensing range	10 cm
Resolution	< 1 mm
Directionality	90°
Response time	Fast – 1 ms
Cost	Low

3.2. Inductive Sensors

An inductive sensor consists of at least two parts: a moving part (known as the target) and a coil, which senses movement of the target by changes in the inductance (L) of the coil. The target is usually a ferromagnetic material such as mild steel. A change in the inductance of the coil is achieved by changing the reluctance of the magnetic circuit of the sensor. Equation 7.4 shows the relationship between inductance and reluctance

$$L = \frac{n^2}{\Re},$$ (7.4)

where n is the number of turns in the coil and $\Re$ is the reluctance of the magnetic circuit.

3.2.1. Magnetic Circuits

A magnetic circuit can be considered to be analogous to an electric circuit. In an electric circuit the electromotive force (*emf*) or voltage (V volts) drives a current (I amps) through a resistor (R ohms), and thus Ohms law can be stated as in equation 7.5:

$$emf = IR.$$ (7.5)

In a magnetic circuit the magnetomotive force (*mmf*; units amp-turns) drives a magnetic flux (ϕ; units Webber) through the magnetic circuit's reluctance $\Re$, and thus

$$mmf = \phi \Re.$$ (7.6)

A magnetic circuit can be treated just like an electric circuit. The reluctance of a magnetic circuit resembles resistance in an electric circuit. Reluctance $\Re$ depends upon the length (d) and area (A) of the magnetic material that is carrying the flux, as shown by equation 7.7,

$$\Re = \frac{d}{\mu A},$$ (7.7)

where μ is the permeability of the magnetic material. The permeability μ is the product of μ_0, the permeability of free space ($4\pi \times 10^{-7}$ Hm^{-1}), and μ_r, the relative permeability of the magnetic material. Therefore μ is analogous to conductivity in an electrical circuit.

Figure 7.5 illustrates two magnetic circuits with their respective electrical circuits of a coil and a constant current power supply. The two magnetic circuits have different reluctance values, and as a consequence the electrical circuits would have different inductance values.

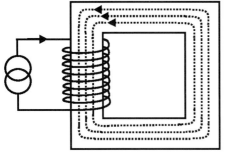
a) Magnetic Circuit of Iron

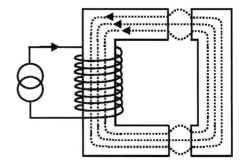
b) Magnetic Circuit of Iron and Air Gaps

Figure 7.5. Magnetic circuits

DISPLACEMENT AND POSITION SENSORS • 181

The magnetic and electrical circuits in figure 7.5 illustrate the magnetic flux lines and electrical current, respectively, showing also the direction of flow. Figure 7.5a shows a magnetic circuit containing only iron; whereas, in figure 7.5b the magnetic circuit contains two air gaps. Assuming that the area and length of the two magnetic circuits are the same, the reluctance of the circuit shown in figure 7.5b will be higher than that shown in figure 7.5a, since air has a much lower permeability than iron.

Therefore the characteristics of the circuits in figure 7.5 are as follows:
- Figure 7.5a (no air gaps) has higher μ → lower $\mathfrak{R}$ → higher L by comparison.
- Figure 7.5b (with two air gaps) has lower μ → higher $\mathfrak{R}$ → lower L.

In an inductive sensor the reluctance of its magnetic circuit can be changed in several ways, including
- varying the length (d) of the magnetic circuit, or
- changing the permeability by moving the magnetic material relative to the coil in the sensor.

Figure 7.6 illustrates these methods. A special example is given by Jansseune [5].

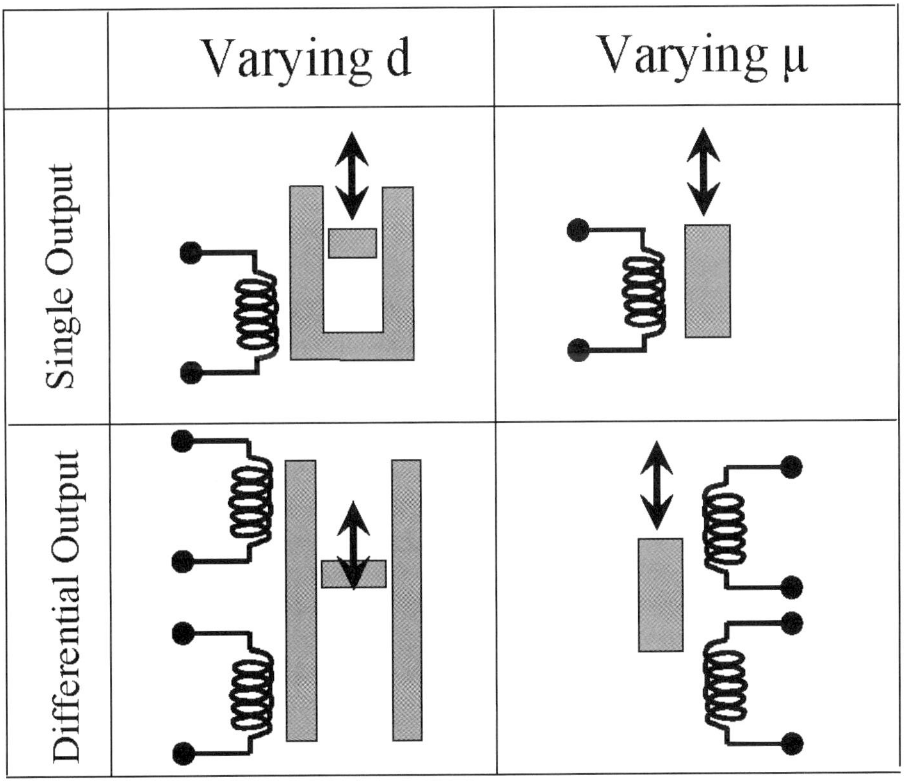

Figure 7.6. General types of inductive sensors

3.2.2. The LDT

The linear displacement transducer (LDT) consists of two coils and a ferromagnetic plunger. The arrangement is illustrated in figure 7.7a. Each coil and the plunger have the same length (s). As the plunger is moved, the inductance of the coils varies. A typical connection configuration is shown in figure 7.7b where the two coils (L_1 and L_2) are connected in a bridge circuit with two resistors (R). An amplifier is also used.

If the inductances are both L when the plunger is in the central position, a displacement of Δs will produce opposing inductance changes of $-\Delta L$ and $+\Delta L$ in the respective coils. Therefore in the ideal case, equation 7.8 is expected as

$$\frac{\Delta L}{L} = \frac{\Delta s}{s} \qquad (7.8)$$

And therefore the corresponding bridge output is given by

$$e_o = \left(\frac{e_i}{2}\right)\left(\frac{\Delta L}{L}\right) \qquad (7.9)$$

However, this ideal case is only true over a limited displacement distance when the plunger is at the center between the two coils.

3.2.3. The LVDT

The linear variable differential transformer is an improvement on the LDT. The device consists of a ferromagnetic plunger, a primary coil, and two secondary coils arranged to measure either linear or

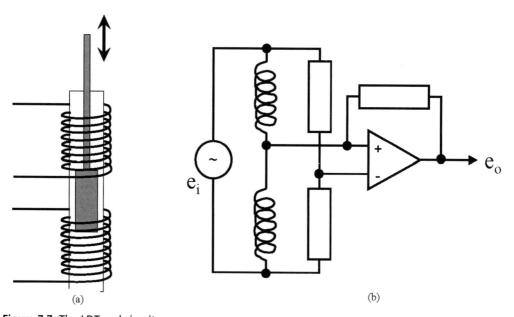

Figure 7.7. The LDT and circuit

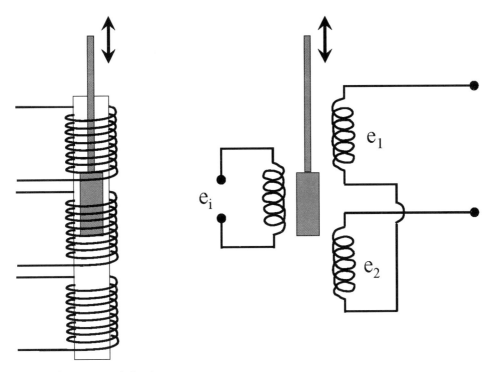

Figure 7.8. The LVDT and circuit

angular motion. Figure 7.8 shows an LVDT arranged to measure linear motion. The movable plunger varies the coupling between the primary and secondary coils.

The primary coil is supplied with an a.c. excitation input e_i, which induces voltages across each of the secondary coils, e_1 and e_2. Normally the secondary coils are connected in series opposition such that an output signal is given as $e_o = e_1 - e_2$. When the plunger is in the central position, the output signal $e_o = 0$. A displacement in either sense from the central position increases the coupling between the primary coil and one of the secondary coils, while at the same time decreasing the coupling with the other secondary coil. As a result, the amplitude of the output signal e_o is a linear function of displacement for most of the sensor range (for 99% of travel range).

Typically an LVDT is supplied with 12–24 V in amplitude at a frequency between 50 Hz to 25 kHz. These transducers may be obtained for travel ranges from ±0.25 mm to ±500 mm. The sensor's usable frequency and travel range are mainly limited by the inertia of the moving plunger and connecting part. These devices have been precluded from production vehicles due to their cost, but are frequently used in prototyping and development studies.

3.2.4. The Variable-reluctance Sensor

The variable-reluctance sensor is the simplest version of this sensor type and will be discussed in more detail in the later section on antilock braking. The sensor is normally used to determine speed of rotation [2,6].

A summary of inductive sensor characteristics is given in table 7.2.

Table 7.2. Inductive sensor summary characteristics

Parameter	Characteristic
Sensing range	10 cm
Resolution	1 mm
Directionality	90°
Response time	Fast – 1 ms
Cost	Low

3.3. CAPACITIVE SENSORS

A capacitive sensor consists of at least two parallel conducting plates. Movement is detected by changes in capacitance resulting from an alteration in geometry or permittivity.

The voltage V across a pair of parallel plates with charge Q is given by equation 7.10:

$$V = \frac{Q}{C}. \qquad (7.10)$$

C is the capacitance of the device in Farads (F). The capacitance can be modified in several ways as illustrated by figure 7.9. Either the distance between the plates (d), the area of the plates (A), or the permittivity of the material between the plates (ε) can be changed. The capacitance is related to these parameters by equation 7.11:

$$C = \frac{d}{\varepsilon A}. \qquad (7.11)$$

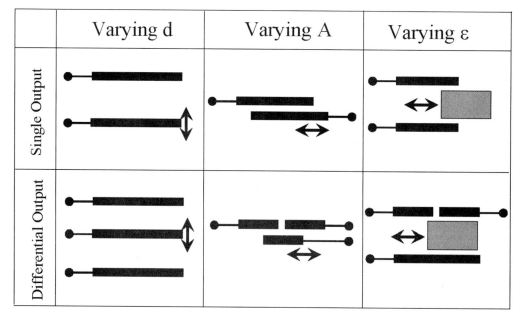

Figure 7.9. General types of capacitive sensors

The permittivity ε of the material between the plates (the dielectric) is the product of the permittivity of free place ε_0 ($\varepsilon_0 = 1/(\mu_0 c^2) = 8.85 \times 10^{-12}$ Fm^{-1}), and the relative permittivity of the material ε_r.

Figure 7.9 illustrates methods for varying the capacitance of a sensor connected for a single output and a differential output. Capacitive sensors are usually employed where accurate measurement of a small displacement is required, such as in pressure or acceleration sensors.

A summary of capacitive sensor characteristics is given in table 7.3.

Table 7.3. Capacitive sensor summary characteristics [7]

Parameter	Characteristic
Sensing range	2 m
Resolution	1 cm
Directionality	90°
Response time	Fast – 1 ms
Cost	Low

3.4. OPTICAL SENSORS

An optical sensor normally consists of three parts: a light source, a light detector, and a target component, which is moved between the light source and detector. This target component modifies the light reaching the detector.

Translational linear and angular optical sensors usually use pulse-counting techniques to monitor movement.

Optical sensors are attractive to the automotive industry because they can be low cost, are made from plastic optical devices, and have high durability since no mechanical connection is made with the sensing component. However, their reliability can be a problem in dirty environments, so their use tends to be limited to cleaner applications, such as steering-wheel position detection [8,9].

3.4.1. Translational Linear Optical Sensors

A translational linear optical sensor moves a target linearly between source and detector systems. A high-precision version uses two diffraction gratings. A fixed grating (known as the scale grating) is stationary. The other (known as the index grating) is fixed to the moving target and slides across the scale grating. This causes the intensity of the light being detected to alternate between a maximum and minimum value. The size of light source and detector do not have to be comparable with the line-distance on the grating. There are, however, advantages in having both of these much larger than the lines on the grating, as any imperfections in the grating are then averaged out.

A single detector system of the type described in the previous paragraph cannot sense the direction of movement. If direction is required a two-detector system must be used with a two-phase index grating as illustrated in figure 7.10. The diagram shows a single light source, an index diffraction grating, a two-phase index diffraction grating, and two detectors, which monitor the light through each phase of the index grating. The two-phase index diffraction grating consists of two tracks of grating lines displaced by one-quarter of a line spacing. The signals from the detectors are then one-quarter out of phase with each other, and thus direction information can be obtained.

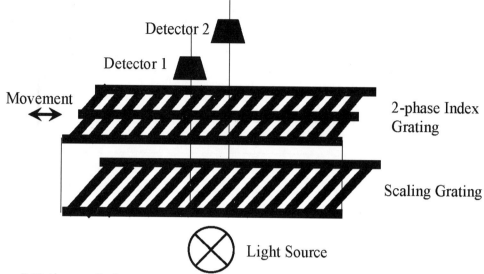

Figure 7.10. Linear optical sensor

Absolute translational linear optical sensors are also available, which use multiple tracks and detectors. The higher the number of tracks and detectors, the higher the precision achieved.

3.4.2. Angular Optical Sensors

These transducers consist of three parts like translational linear sensors. A light source, a detector, and a rotatable disk are used. The disk can be slotted or may have reflective strips that "chop" or reflect the light.

If a disk carries a series of even, equally spaced slots, the pulses from the detector can indicate rotational speed. For N revolutions per minute, the output pulse rate R is given by

$$R = \frac{NS}{60} Hz, \quad (7.12)$$

where S is the number of slots in the disk.

An incremental angular encoder consists of a slotted disk, as described above, and cannot give absolute position information unless the starting point is known.

An absolute angular encoder is used if both the absolute position and direction of travel are required. Figure 7.11a shows the disk from an absolute angular encoder, with four sets of slots making it a four-bit encoder. This four-bit example can detect 16 different angular positions, with a resolution of 30°. For higher resolution more bits (b) are required as described by equation 7.13 with the angle of resolution θ.

$$\theta = \frac{360}{2^b} \quad (7.13)$$

Figure 7.11a shows a disk with natural binary code. Figure 7.11b is a cross section of the disk showing the position of the four light sources and four detectors. Figure 7.11c is a disk with Gray code, which is used to minimize errors in position determination.

In Gray code only one bit at a time changes as the disk rotates.

DISPLACEMENT AND POSITION SENSORS • 187

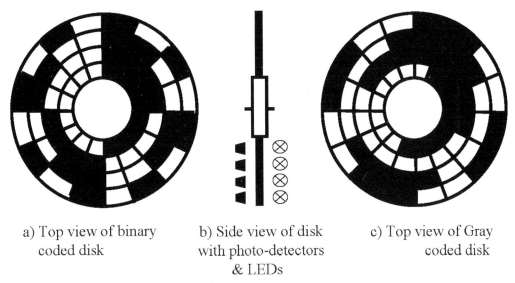

Figure 7.11. Angular encoder for optical sensor

The disk arrangements discussed so far use a parallel configuration of detectors and slotted tracks. An alternative further arrangement is to configure the detectors in series with just one track of slots. This saves on space and material.

Another alternative is to use the angular optical position sensor developed by using a spiral slit [10]. This gives a linear output analogue voltage signal.

A summary of the optical sensor characteristics is given in table 7.4.

Table 7.4. Optical Sensor Summary Characteristics [6]

Parameter	Characteristic
Sensing range	100 m
Resolution	Low
Directionality	Good
Response time	Medium 100 ms
Cost	High

3.5. RADAR-BASED SENSORS

*RA*dio *D*etection *A*nd *R*anging (radar) systems use electromagnetic waves to detect a target and measure its distance and direction. A pulse of waves is transmitted, which reflects off a target, and the return is received at the detector. The time taken from emitting the wave to receiving the reflection is used to determine the distance of the target. Figure 7.12 shows the arrangement.

Radar-based sensors have been developed for several automotive applications. A forward-recognition sensor that uses radar has been developed for use in a traffic congestion support system [11]. The

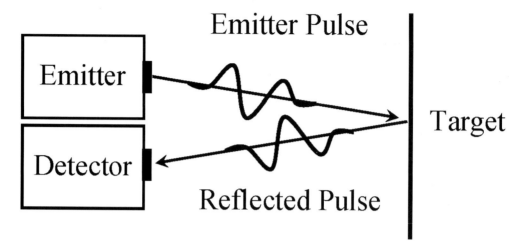

Figure 7.12. Radar sensor

system is used to minimize rear-end collisions on congested roads. Active cruise control (ACC) systems use radar-based sensors [12] along with near distance sensing [13].

A summary of radar sensor characteristics is given in table 7.5.

Table 7.5. Radar sensor summary characteristics [6]

Parameter	Characteristic
Sensing range	150 m
Resolution	10 mm
Directionality	2°
Response time	Fast – 1 ms
Cost	High

3.6. Ultrasonic Sensors

Ultrasonic sensors use sound waves (ultrasound) at frequencies above 18 kHz (higher than the human ear can detect). Ultrasound technologies have been used for many applications, such as marine sonar (*SO*und *N*avigation *A*nd *R*anging), medical ultrasound imaging, and structural defect location systems. These systems use sound echoes to detect a target. An emitter transmits a series of pulses of sound waves, which reflect off the target, and are received at the detector. The time taken from emitting the wave to receiving the reflection is used to determine the distance of the target.

L. F. Richardson [14] was the first around 1912 to suggest the use of an ultrasonic echo detection system. This idea was developed during the First World War for the detection of submarines, with the first practical device build by P. Langevin. Modern marine sonar systems have developed from this initial work.

In automotive applications ultrasound has been used to measure the ride height of a vehicle for adaptive-suspension control. These ultrasound sensors have also been developed for sensing passenger and driver head positions for improved safety [15]. A summary of ultrasonic sensor characteristics is given in table 7.6.

Table 7.6. Ultrasonic sensor summary characteristics [6]

Parameter	Characteristic
Sensing range	< 10 m
Resolution	10 mm
Directionality	30°
Response time	60 ms
Cost	Medium

3.7. Hall Effect Sensors

An electric current (I) is carried by moving charged particles in a metal or semiconductor. If the current-carrying material is placed in magnetic field (B) normal to the flow, the moving charged particles experience a force perpendicular to both the flow of current and magnetic field. This is known as the Lorentz force, and it causes charged particles to accumulate at the two sides of the material. Positive charges accumulate on one side and negative charges accumulate at the other, producing a charge

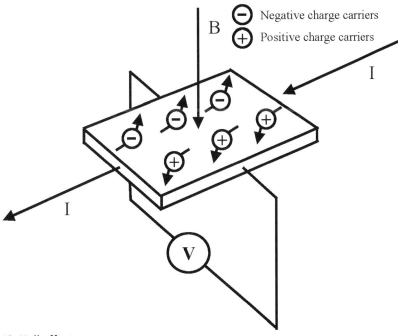

Figure 7.13. Hall effect

imbalance which can be detected with a voltmeter; see figure 7.13. This phenomenon is known as the Hall effect, and it can be utilized for a number of sensor applications [16].

For a slab of semiconductor with thickness h, the Hall voltage V developed across the width of the material is given by equation 7.14,

$$V = \frac{K_H BI}{h}, \qquad (7.14)$$

where K_H is the Hall coefficient of the material. K_H depends on the charge mobility and resistivity of the material. Iridium antimonide is often used in Hall devices and has $K_H = 20\ VT^{-1}$.

Hall effect sensors can be used for noninvasive current measurement, magnetic field measurement, and proximity sensing. A Hall effect rotary position sensor has been used for throttle position measurement [17].

A summary of Hall effect sensor characteristics is given in table 7.7.

Table 7.7. Hall effect sensor summary characteristics

Parameter	Characteristic
Sensing range	Medium – 1–10 mm
Resolution	High – < 1 mm
Directionality	90°
Response time	Fast
Cost	Medium

3.8. MISCELLANEOUS SENSORS

Sensor technologies are continually being developed [2]. This section concludes with a discussion of one of the simplest and cheapest forms of sensor, the switch or microswitch. Switches are used in many sensing applications, for example, to indicate when a door is opened.

A summary of switch sensor characteristics is given in table 7.8.

Table 7.8. Switch type summary characteristics

Parameter	Characteristic
Sensing range	Two states only—On/Off
Resolution	n/a
Directionality	n/a
Response time	Medium –10 ms and slower
Cost	Low

4. MEASURANDS IN THE AUTOMOTIVE APPLICATION

This section gives an overview of automotive measurands based on position, displacement, or velocity sensors. The complexity of automotive control and monitoring systems has grown substantially in the 100-year life of the modern road vehicle. Thus the need for sensing has also grown.

4.1. MEASURANDS BASED ON DISPLACEMENT OR POSITION

4.1.1. Liquid Level Detection

The level of fuel in the fuel tank is normally detected by a position sensor. A potentiometric displacement sensor connected to a float and level arm is one of the simplest designs, and is a cost-effective solution.

Similar systems can be adopted for other liquid reservoirs, such as water coolant, brake fluid, and other hydraulic fluid levels.

4.1.2. Proximity Detection

Cabin lighting activated when a door is opened is one of the simplest sensors and uses a switch arrangement.

4.1.3. End-of-Travel Detection

Switch arrangements can also be used as an end-of-travel sensor for electric windows and for positioning of, for example, mirrors and chairs.

4.2. ENGINE MANAGEMENT SYSTEMS

The engine management system is one of the most complex control systems in the motor vehicle [18]. The system is designed to keep the engine operating at optimum performance for different load conditions. Some of the important measurands required to achieve this are airflow rate, throttle position, engine speed, and the manifold pressure.

4.2.1. Airflow Rate

The vane sensor (see also chapter 4) is an airflow sensor in which a spring-loaded vane is deflected by the airflow. Higher airflow gives a larger deflection of the vane, and vane deflection is normally detected by a potentiometer.

4.2.2. Throttle Position

Throttle position is normally measured by a potentiometer in order to turn the deflection into an electrical signal [17].

4.2.3. Engine Speed

The engine speed sensor, which monitors the crankshaft, is usually an inductive device that senses the passing of teeth on a gearwheel. As the wheel rotates, the teeth pass the sensor changing its inductance. The changes in inductance can be used to determine the speed of the engine in revolutions per minute (RPM) [19].

4.2.4. Manifold Pressure

A diaphragm deflection caused by the pressure being measured can be connected to a potentiometer arrangement to change the deflection into an electrical signal. See chapter 2 for more details.

4.3. VEHICLE CONTROL SYSTEMS

Vehicle control systems for ensuring a smooth and fuel-efficient ride are growing in importance. A few of the most important such systems are highlighted in this section.

4.3.1. Braking System Sensors

A wheel speed sensor may be used to determine whether the wheel has "locked" during braking. This will be discussed in greater detail in a subsequent section.

4.3.2. Suspension System Sensors

An active suspension system enables virtually constant wheel-load and constant mean vehicle height. A position sensor may be used to determine the correct height of the vehicle.

4.3.3. Vehicle Dynamic Control Sensors

A vehicle dynamic control (VDC) system is designed to present lateral instability of the vehicle. It is integrated with the braking system and traction control to prevent the vehicle from "pushing out" of a turn or spinning out of the turn. Typical measurands used are wheel speed, braking pressure, steering position [20], yaw rate, and lateral acceleration [21].

4.4. SAFETY SYSTEMS

The number of safety systems in a road vehicle continues to increase. These systems are becoming ever more complex and involve many sensors [22,23].

4.4.1. Seatbelt Sensors

One of the simplest safety systems is to use a warning chime, reminding the occupant to put on the seatbelt (although many have reported that this is one of the most annoying features of the modern vehicle!). This requires a system to detect the occupant in the seat: The normal solution is to use a microswitch.

4.4.2. Airbag Sensors

Acceleration sensors are normally used to trigger the deployment of an airbag when a crash occurs. More advanced airbag systems also take account of the occupant's position. Position sensors have been used to detect the position of the occupant to maximize safety. An occupant detection system would be used to determine whether an infant is in a child seat and thus would inflate the appropriate airbag only if necessary [24]. These systems can also be used to detect where a passenger is out of position, again to maximize safety if an airbag needs to be inflated.

4.4.3. Rollover Protection Sensors

Advanced rollover sensors are used to detect that the vehicle has turned over and thus deploy a roll bar to protect the occupants. Commercial sensor units continuously monitor several parameters, such as the angle of the vehicle and the lateral and vertical acceleration.

4.5. MISCELLANEOUS SYSTEMS

4.5.1. Parking Assist Systems

A parking assist system uses between 4 and 10 distance sensors or rangefinders, depending on the sophistication required. These rangefinders are normally ultrasonic and are fitted to the front and rear of a vehicle to detect any obstacles when the driver is parking the vehicle.

5. EXAMPLE OF AUTOMOTIVE SENSOR SYSTEM APPLICATION

Antilock braking systems (ABS) are commonly fitted to passenger vehicles. The objective of such a system is to optimize braking and improve safety. The braking performance of the modern vehicle has been improved greatly with sophisticated electronic components. Some doubt, however, whether the full performance of the ABS is utilized [25].

This section looks in detail at antilock braking systems and the use of the variable-reluctance and other sensors within ABS.

5.1. OVERVIEW OF THE BRAKING SYSTEM

Road vehicles have two braking systems [2]:

- The parking brake
- The service brake

5.1.1. Service Braking System

The service braking system allows the driver to reduce vehicle speed during normal operation and allow the vehicle to be brought to a stationary position.

5.1.2. Parking Braking System

The parking brake allows the vehicle to be held in a stationary position in the absence of the driver. The system is designed to keep the vehicle stationary on a steeply inclined surface.

5.1.3. Antilock Braking System (ABS)

This system involves the service braking system and allows for automatic adhesion-slip-rate control for all wheels. The system stops the wheels from locking under full braking conditions and allows optimum braking of the vehicle.

5.1.4. The Operation of the Braking System

The braking system of a vehicle is designed to dissipate the kinetic energy of the motion of the vehicle as heat. Friction brake pads and disks are used on each wheel to achieve this.

Figure 7.14 illustrates a typical configuration of brake pads and disk. The brake disk is fitted to the wheel of the vehicle. The hydraulic actuator (a caliper with piston and pads) is fixed to the chassis of the vehicle. When the brake is operated, hydraulic fluid is forced into the piston on the caliper, which squeezes the brake pads onto the disk causing it to grip the disk and slow the rotation of the wheel.

A "muscular-energy braking system" is a braking system where the hydraulic fluid is forced into the pistons solely by the physical effort of the driver; that is, the braking force is solely provided by the driver. Most modern vehicles use an "energy-assisted braking system," that is, a braking system in which the hydraulic fluid is forced into the actuator pistons by a combination of the driver's effort and another energy source. The normal method of assisting the driver in a passenger vehicle is to use a vacuum-operated brake booster (also known as a servo-assisted brake).

Figure 7.15 is a plan view of the layout of the four wheels of the vehicle for two of the most common configurations of hydraulic circuits; known as the II distribution and the X distribution.

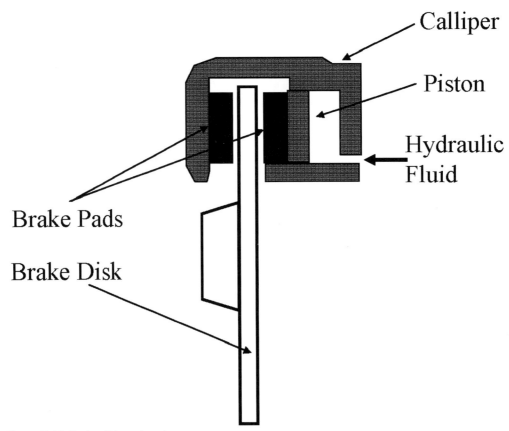

Figure 7.14. Brake disk and pads

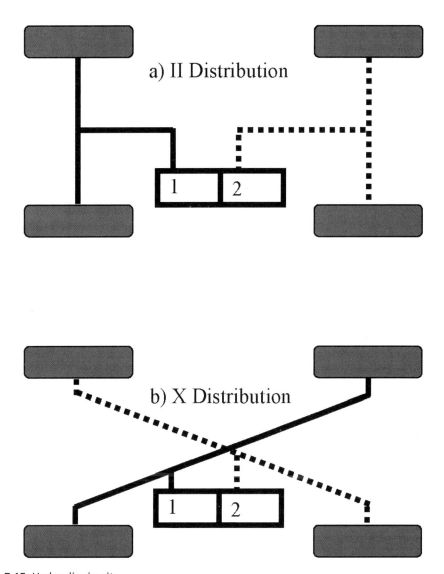

Figure 7.15. Hydraulic circuits

5.2. ADHESION-SLIP CHARACTERISTIC

When the brake pedal is pressed, hydraulic fluid pressure rises and causes the pads to clamp the brake disks. If the pedal is pressed hard enough the wheels will lock. This occurs when the brake slip λ rises beyond the maximum point on the adhesion-slip curve (figure 7.16) and moves from the stable to the unstable region. Brake slip λ is defined as the ratio of the difference between the wheel speed v_r and the vehicle speed v_f, divided by vehicle speed and expressed as a percentage,

$$\lambda = \frac{(v_f - v_r)}{v_r} \cdot 100 \quad (7.15)$$

where the wheel speed is given by

$$v_r = \omega r \quad (7.16)$$

where ω is the angular velocity of the wheel and r is the radius of the wheel.

Figure 7.16 is a plot of three conditions for the adhesion-slip curve.

Figure 7.17 shows the forces acting on a wheel. G is the downward force due to gravity G, and F_B is the braking force,

$$F_B = \mu_B G \quad (7.17)$$

where μ_B is the braking force coefficient.

In the stable region of the adhesion-slip curve (figure 7.16), brake slip largely arises through deformation of the tire. This is known as deformation slip. The braking becomes increasingly "skidding" in the unstable region of the curve.

The braking force coefficient μ_B drops sharply once the peak of the graph is passed and the unstable region is reached. The resulting excess torque causes the wheel to brake to a stop very rapidly

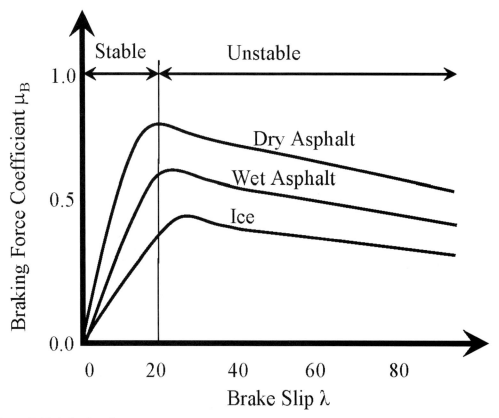

Figure 7.16. Adhesive-slip curve

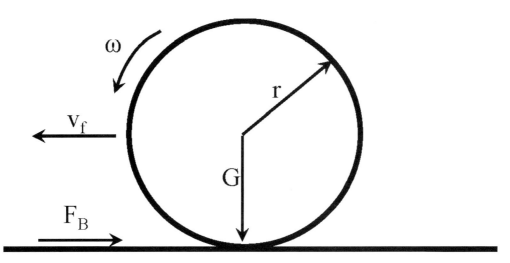

Figure 7.17. Schematic diagram of forces

in the absence of ABS, which normally results in a skid. The effect can be detected by monitoring the motion of the wheel using a speed sensor.

5.3. MEASURING SLIP

The most straightforward way to control tire slip in an ABS would be to measure the vehicle speed v_f and the wheel speed v_r directly,

$$v_r = \frac{2\pi r N}{60} = 2\pi r f, \qquad (7.18)$$

where N is the number of rotations per minute (RPM) or f is the frequency of rotation.

Slip happens when vehicle speed exceeds the wheel speed, that is

$$v_f > v_r. \qquad (7.19)$$

However, there are at present no cost-effective vehicle speed sensors that do not depend on measuring the speed of a wheel. However, an effective ABS system can be designed without a direct measurement of v_f. The changes in vehicle speed that can occur without causing slip are limited by the inertia of the vehicle. However, the wheel speed can change very rapidly when tire slip occurs. Therefore, the wheel angular acceleration (or deceleration) is much higher when slip occurs. The angular acceleration ω' is readily estimated from the rate of change of wheel speed

$$\omega' = \frac{v_r(t_1) - v_r(t_2)}{t_1 - t_2}, \qquad (7.20)$$

where $v_r(t_1)$ and $v_r(t_2)$ are the wheel speeds at time internal t_1 and t_2. The tire slip corresponds to the angular acceleration exceeding a threshold value A:

$$\begin{aligned} A > \omega' &\rightarrow slip \\ A \leq \omega' &\rightarrow no\ slip \end{aligned}. \qquad (7.21)$$

5.4. THE VARIABLE-RELUCTANCE SENSOR

A variable-reluctance sensor is often used to monitor the speed of a wheel v_r. A sensor is normally fitted to each wheel.

Figure 7.18 shows a schematic diagram of a variable-reluctance sensor for two conditions. Each diagram illustrates the sensor, which consists of an electromagnet and a target consisting of a ferrous metal wheel with teeth. As the wheel rotates there are two possible conditions: either alignment of the teeth with the sensor or misalignment. Under condition "a" (when there is a maximum air gap) the reluctance of the magnetic circuit will be high and thus inductance of the electrical circuit will be low. Under condition "b" (a minimum air gap) the reluctance of the magnetic circuit will be low and the inductance of the electrical circuit will be high.

Changes in the inductance of the electrical circuit appear as voltage pulses as the wheel rotates.

5.5. THE ANTILOCK BRAKING SYSTEM

The main electronic component of an ABS consists of two counters, which alternately measure the wheel speed. The counters are switched many times a second. The wheel speed is estimated from the sensor pulse rate.

Figure 7.19 is a schematic diagram illustrating a typical arrangement of the electronic components.

Wheel deceleration (or acceleration) is measured by testing whether the earlier wheel speed exceeds the later wheel speed by a preset threshold value [26]. When this occurs a skid condition has been detected. The ABS then generates an electrical signal, which reduces hydraulic pressure by means of a solenoid-actuated relief valve known as a modulator. The brake pressure is reduced by an amount sufficient to eliminate the locked brake.

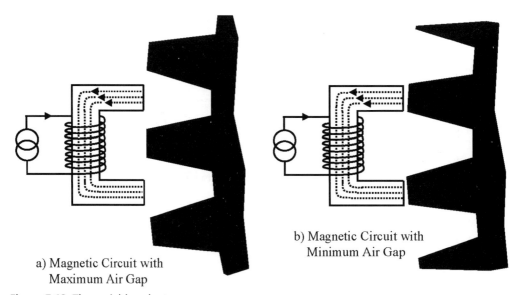

a) Magnetic Circuit with Maximum Air Gap

b) Magnetic Circuit with Minimum Air Gap

Figure 7.18. The variable-reluctance sensor

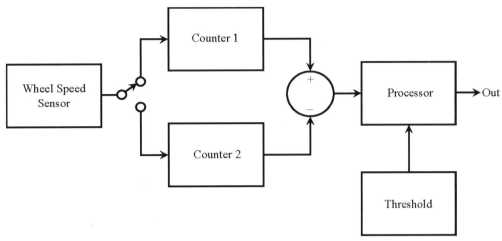

Figure 7.19. Schematic diagram of the ABS electronic components

5.6. EXAMPLE CALCULATION

5.6.1. The Question

In an antilock braking system based on wheel speed measurement, a variable-reluctance wheel speed sensor is used that provides 500 pulses per wheel revolution. The wheel radius is 300 mm. Initially (before skidding) the tangential tire speed is 25 ms^{-1}. One-tenth of a second later it is found to be 5 ms^{-1}. Estimate the brake slip λ after 0.1 seconds and state whether the braking is stable or unstable. If the limiting value of the stable braking is a brake slip of 22%, what is the threshold value for the ABS intervention with the vehicle in the condition described above?

5.6.2. The Solution

From data given, the vehicle speed (which initially equals the tangential tire speed) $v_f = 25$ m/s

Assuming the vehicle speed is unchanged 0.1s later, but that the tire is skidding ($v_r = 5$ m/s), the brake slip λ can be found from equation 7.21 and is given in equation 7.22 for this data:

$$\lambda = \left(\frac{25-5}{25}\right) \times 100 = 80\% \qquad (7.22)$$

Clearly λ is in the unstable region of the brake slip λ curve.

If the limit for stability is $\lambda = 22\%$ then at a vehicle speed of 25 m/s the threshold v_r is 19.5 m/s (from equation 7.21). If the ABS sensor returns 500 pulses per revolution, the threshold pulse rate is found from equation 7.23:

$$v_r = 19.5 \, \omega r$$
$$\omega = 19.5/0.3 = 65 \; rad/sec \equiv 10.35 \; rps, \qquad (7.23)$$

where $r = 300$ mm.

Therefore, at 500 pulse/rev, the threshold pulse rate is 5172 Hz.

6. FUTURE PERSPECTIVES

The need for position and displacement sensors in automotive engineering continues to rise.

6.1. NOVEL SENSING DEVICES

Sensor developments in the future are likely to be concerned with improving reliability, increased accuracy, ease of implementation, and reduced costs. These aims may be achieved by the use of new and improved materials and better manufacturing techniques.

Novel materials (such as the magnetoresistive semiconductors) have been used in rotary sensors. These give high accuracy at very low speeds [27–29].

Microelectromechanical systems (MEMS) technology is slowly replacing the more conventional discrete sensor in many automotive applications. Airbag sensors and air conditioning are current examples where this technology is used. MEMS are micron-sized electrical and mechanical components on a single silicon wafer and manufactured in a similar process to the integrated circuit on a chip.

6.2. EMERGING VEHICLE TECHNOLOGIES

Intelligent transport systems (ITS) are under development to optimize traffic flow, reduce pollution, and improve safety [30]. ITS systems involve vehicle-to-vehicle communication [31], as well as vehicle-to-infrastructure communication.

New and improved safety systems are the subject of much research and development effort.

7. SUMMARY

Displacement and position sensors for automotive applications are employed across a wide range of systems in the passenger vehicle. Many technologies are used, from low-cost potentiometric and inductive sensors to high-cost radar-based devices. These sensors are used to monitor linear translations or angular movement in systems such as engine management, vehicle control, safety, and so forth.

The antilock braking system (ABS) is one of the most common systems fitted to passenger vehicles. The objective of such a system is to optimize braking and improve safety. The braking performance of the modern vehicle has been improved greatly with sophisticated electronic components. The variable-reluctance sensor is a typical device used on an ABS for monitoring the speed of each wheel to which it is fitted. Changes in speed of the wheel beyond a particular threshold under the braking condition indicate that the wheel is entering the "locked" state. This state will result in the vehicle skidding unless intervention occurs, so the ABS releases the brake fluid pressure slightly to prevent the skid.

NOMENCLATURE

A Area (in m^2) or Threshold value
B Magnetic flux density/T
C Capacitance/F
d Length/m

e	Potential difference/V
F_B	Braking force/N
G	Force due to gravity/N
h	Thickness of material/m
H	Heat dissipation/W
I	Electrical current/A
L	Inductance/H
K_H	Hall coefficient/VT^{-1}
n	Number of turns on a coil
N	Revolution per minute/RPM
Q	Electrical charge/C
r	Radius/m
R	Electrical resistance/Ω or speed of rotation/Hz
S	Number of slots
t	Time/s
v	Velocity/ms^{-1}
V	Voltage/V
x	Position/m
$\mathfrak{R}$	Reluctance/AWb^{-1}
emf	Electromotive force/V
mmf	Magnetomotive force/A-turns
λ	Brake slip Lambda
ε	Permittivity/Fm^{-1}
θ	Angle/deg or rad
φ	Magnetic flux/Wb
μ	Permeability/Hm^{-1}
μ_B	Braking force coefficient
ω	Angular velocity/rad s^{-1}

REFERENCES

1. Bauer, H., ed. 1996. *Automotive handbook*. Warrendale, PA: SAE.
2. Jurgen, R. K. 1999. *Automotive electronics handbook*. 2nd ed. New York: McGraw-Hill.
3. Westbrook, M. H., and J. D. Turner. 1994. *Automotive sensors*. Bristol: IOPP.
4. Atkinson, J. K. 1989. *Transducers, principles and practice*. London: NEMEC.
5. Jansseune, L., B. Legrand, J. Y. Voyant, and J. P. Yonnet. 2003. Planar-coil displacement sensors. In *Sensors and Transducers,* ed. R. K. Jurgen. Detroit: SAE.
6. Jawad, B., N. Hachem, W. Bowerman, J. Leese, and S. Cizmic. 2003. Engine control inputs and signal conditioning for crankshaft and camshaft positioning. In *Sensors and Transducers Pt. 105,* 189–92. 2nd ed. Detroit: SAE.
7. Upton, M. 1997. Techniques for distance measurement. In *Sensors and Transducers Pt. 68,* 63–69. Detroit: SAE.
8. Goodyer, E. N. 1989. An overview of a range of novel automotive sensors. In *Proceedings of 7th Int. Conference on Automotive Electronics,* 79–86. London: Institution of Mechanical Engineers.
9. Smith, D. S., and P. R. Jackman. 1992. Optical Sensors for Automotive applications. In *IEE Computing and Control Division Coll. 1997/107 on Automotive Sensors,* 2/1–2/3. London: IEE.

10. Kato, S. and J. Nakaho. 1997. High precision angular position sensor. In *Sensors and Transducers Pt. 68*, 63–69. Detroit: SAE.
11. Higashida, H., R. Nakamura, M. Hitotsuya, K. F. Honda, and N. Shima. 2003. Fusion sensor for an assist system for low speed in traffic congestion using millimeter-wave radar and an image recognition sensor. In *Sensors and Transducers*. 2nd ed., pt. 105: 171–75.
12. Prestl W., T. Sauer, J. Steinle, and O. Tschernoster. 2003. The BMW active cruise control ACC. In *Sensors and Transducers Pt. 105*, 205–11. 2nd ed. Detroit: SAE.
13. Kunert, M. 2003. Radar-based near distance sensing device for automotive applications. In *Sensors and Transducers Pt. 105*, 213–19. 2nd ed. Detroit: SAE.
14. Turner, J. D., and A. J. Pretove. 1991. *Acoustics for engineers*. Basingstoke: Macmillan.
15. Massara, A. J. 1997. Ultrasonic sensing of head position for head restraint automotive adjustment. In *Sensors and Transducers Pt. 68*, 71–74. Detroit: SAE.
16. Goodwin, W. C., J. Gilmore, and J. C. Rustman. Math based approach to model active Hall Effect position sensors. In *Sensors and Transducers Pt. 105*, 177–83. 2nd ed. Detroit: SAE.
17. Bicking R., G. Wu, J. Murdock, D. Hoy, and R. Johnson. R. 1997. A Hall Effect rotary position sensor. In *Sensors and Transducers Pt. 68*, 81–83. Detroit: SAE.
18. Washino, S. 1988. *Japanese technology reviews: Automobile electronics*. Vol. 1. Melbourne: Gordon and Breach Science Publishers.
19. Yoshino, Y., K. Ao, M. Kato, and S. Mizutani. 1997. MRE rotation sensor: High-accuracy, high-sensitivity magnetic sensor for automotive use. In *Sensors and Transducers Pt. 68*, 85–91. Detroit: SAE.
20. Gruber, J. 1997. Steering wheel angle sensor for vehicle dynamics control systems. In *Sensors and Transducers Pt. 68*, 57–61. Detroit: SAE.
21. Kikuchi T., Y. Osugi, M. Tani, Y. Iwata, S. Gouji, S. Ishikawa, T. Enokijima, S. Yokoi, and T. Soma. 2003. Flat quartz angular rate sensor using hammer-headed double-T structure. In *Sensors and Transducers Pt. 105*, 167–70. 2nd ed. Detroit: SAE.
22. Hagleitner, W. and F. Bernal. 2003. Lane detection: A new, low cost system offering several unique application opportunities. LIS100. In *Sensors and Transducers Pt. 105*, 201–3. 2nd ed. Detroit: SAE
23. Hwang, J. K., and C. K. Song. 2005. Fuzzy estimation of vehicle speed using an accelerometer and wheel sensors. *International Automotive Technology* 6(4): 359–65.
24. Steiner, P., and G. Wetzel. 2003. New aspects on static passenger and child seat recognition and future dynamic out-of-position detection for airbag control systems. In *Sensors and Transducers Pt. 105*, 233–37. 2nd ed. Detroit: SAE.
25. Park, S. 2005. Effect of brake pedal impedance on braking performance in EH-BBW system. *International Automotive Technology* 6(4): 391–402.
26. Ivanov, V., M. Belous, S. Liakhau, and D. Miranovich. 2005. Results of functional simulation for ABS with pre-extreme control. *International Automotive Technology* 6(1): 37–44.
27. Ishiai Y., N. Jitousho, T. Korechika, J. LeGare, and S. Yoshida. 1997. High accuracy semiconductive magnetoresistive rotational position sensor. In *Sensors and Transducers Pt. 68*, 47–55. Detroit: SAE.
28. Taguchi M., I. Shinjo, T. Kawano, M. Ikeuchi, and Y. Ohashi. 2003. GMR revolution sensors for automobiles. In *Sensors and Transducers Pt. 105*, 185–88. 2nd ed. Detroit: SAE.
29. Weser, M., and A. Harmansa. 2003. Novel self-monitoring magnetoresistive sensor system for automotive angular measurement applications. In *Sensors and Transducers Pt. 105*, 159–66. 2nd ed. Detroit: SAE.
30. Tanaka, S., T. Nakagawa, M. Akasu, Y. Fukushima, and W. Bracken. 2003. Development of a compact scan laser radar. In *Sensors and Transducers Pt. 105*, 227–32. 2nd ed. Detroit: SAE.
31. Wagner, M., J. Dickmann, J. Büchler, V. Winkler, U. Siart, and J. Detlefsen. 2003. Radar based IVC system. In *Sensors and Transducers Pt. 105*, 149–52. 2nd ed. Detroit: SAE.

CHAPTER 8

ACCELEROMETERS

Jonathan Swingler

1. INTRODUCTION

Shock and vibration are present in all areas of our daily lives. Whenever a structure moves it experiences acceleration. The ability to characterize this acceleration leads to a better understanding of the dynamic characteristics that govern the behavior of the structural movement. These measurements are vital for the development, testing, and operation of structures and machines in all fields of engineering. The resulting data can be used to modify response, enhance ruggedness, improve durability, and/or reduce noise and vibration.

An accelerometer is an electromechanical transducer that converts the mechanical motion of a mass into an electrical signal proportional to the acceleration of that mass, that is, the rate of change of velocity with respect to time. Such transducers obey the second law of Newtonian mechanics in that the force acting on the measuring element is directly proportional to the acceleration produced:

$F = mass \times acceleration$, i.e.,

$$F = ma \qquad (8.1)$$

The principles of acceleration sensing can and have been used in a wide variety of applications. Accelerometers can also be configured for inertial measurement of velocity and position, for vibration and shock measurement, and with state-of-the-art accelerometers that incorporate the measurement of gravity, to determine orientation by providing tilt and inclination information. The automotive industry uses many millions of accelerometers. Accelerometers are strategically positioned within a vehicle for analyzing the signature profile of acceleration, velocity, and distance to distinguish a crash from a noncrash event. The decision to fire airbags has to be made within a few tens of milliseconds to protect vehicle occupants. The increasing sophistication of automation in a car means that there are accelerometers that sense the tilt of a car for antitheft systems. Accelerometers may be integrated with GPS systems to provide accurate positional information [1].

A review carried out by Yazdi and others [2] on micromachined inertia sensors showed that accelerometer applications can be found not only in automotive industries, but also in biomedical, consumer, military, and navigation and guidance systems. These applications range from the use of accelerometers in the active stabilization of pictures taken by camcorders, head-mounted displays, safing and arming in missiles and other ordnance, self-contained navigation and guidance systems, seismometry for oil exploration, earthquake prediction, and platform stabilization in space. In the Indianapolis 500, racing circuit drivers wear tiny accelerometers inside custom-made earpieces designed to make racing safer. These earpieces, manufactured by Sensaphonics, contain microelectronic accelerometers that measure the g-forces on drivers' heads during crashes. However, the largest volume demand for accelerometers is probably due to their automotive applications. Recent research confirms that the market for acceleration, vibration, and velocity measurement devices—universally known as accelerometers—is continuing to grow. Frost and Sullivan have projected that future market revenues will exceed U.S. $1 billion, with total growth estimated at 4.7% annually through 2007 [3].

Accelerometers can be classified according to their transduction mechanisms. The most common type is the "piezo" class, which can be further subdivided into piezoresistive or piezoelectric types. The piezoresistive effect occurs when a material changes its electrical resistance in response to an applied stress; whereas, a piezoelectric material—which may be a crystal or a synthetic ceramic material such as PZT—produces a charge separation when subjected to a compressive force.

Capacitive and optical sensing techniques can also be found in accelerometer designs. Accelerometers with capacitive sensing elements typically use a so-called proof mass as one plate of the capacitor and the base as the other. When the sensor is accelerated, the proof mass moves and the voltage across the capacitor changes with the applied acceleration. An optical accelerometer uses a sensor that measures the distance to a membrane by determining the intensity of light reflected from that membrane. These sensing principles will be discussed in more detail in section 3.

In this chapter the design characteristics, operating principles, and limitations of the most common forms of accelerometers are described. Clearly, few, if any, users will wish to design their own accelerometer, but an understanding of how a sensor operates is an essential prerequisite to its intelligent selection and use. Hence, the following section will introduce the user to the fundamental theory of accelerometer operation. Choosing the best form of accelerometer to use in a given situation can be difficult because of the wide selection of sensing principles available. Furthermore, when using an accelerometer (or any other form of sensor) the effect of environmental changes on the output signal must be considered. Changes in the thermal, magnetic, and acoustic environment in which the transducer is situated may all affect the sensor in this way. Accelerometers can also give erroneous outputs because of base strain if incorrectly mounted, from humidity changes, or from electromagnetic interference (EMI).

2. THEORY OF OPERATION

A detailed theory of operation of the accelerometer can be found in reference 20, but in its simplest form, it can be regarded as a seismic mass restrained by a spring and appropriately damped. When acceleration is applied to the accelerometer case, the mass moves relative to the case. When the acceleration stops the spring returns the mass to its original position. Under steady-state acceleration conditions, a seismic mass that experiences a constant acceleration will move from its rest position to a new position as determined by the balance between its mass multiplied by the acceleration and the restoring force of the spring. Using a simple mechanical spring, the acceleration will be directly proportional to the distance traversed by the seismic mass from its equilibrium position.

The dynamics of such a system—depicted in figure 8.1—can be described in simple terms by the second-order system of a so-called mass-spring damper. Transducers of this nature operate by sensing the motion of a suspended mass m relative to the case. The equation of motion for the system in figure 8.1 is as follows:

$$m\ddot{x} = -c(\dot{x} - \dot{y}) - k(x - y) \qquad (8.2)$$

In this equation, the displacement of the seismic mass and the vibrating body is represented by x and y respectively. If the relative displacement z, between the seismic mass and the case is

$$z = x - y, \qquad (8.3)$$

and if the vibrating body to which the case is attached undergoes sinusoidal motion of the form $y = A \sin \omega t$, then equation 8.2 becomes

$$m\ddot{z} + c\dot{z} + kz = m\omega^2 A \sin \omega t \qquad (8.4)$$

This is the well-known equation for the response of a system with a single degree of freedom to forced vibration. The steady-state solution can be assumed to be of the form $z = Z \sin(\omega t - \phi)$ where z is the amplitude of the mass-case displacement and ϕ is the phase of the displacement with respect to the exciting force. By comparison with the standard solutions we can write down equations 8.5 and 8.6:

$$z = \frac{m\omega^2 A}{\sqrt{(k - m\omega^2)^2 + (c\omega)^2}} = \frac{A(\omega/\omega_n)^2}{\sqrt{\left[1 - (\omega/\omega_n)^2\right]^2 + \left[2\zeta(\omega/\omega_n)\right]^2}} \qquad (8.5)$$

$$\tan \varphi = \frac{\omega c}{k - m\omega^2} = \frac{2\zeta(\omega/\omega_n)}{1 - (\omega/\omega_n)^2} \qquad (8.6)$$

These equations show that the important parameters are the frequency ratio ω/ω_n and the damping factor ζ and are plotted in figure 8.2. The type of sensor is determined by the relationship between the frequency to be measured and the resonance frequency of the transducer.

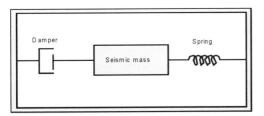
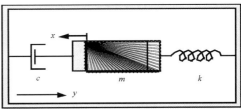

Figure 8.1. A damped mass-spring system

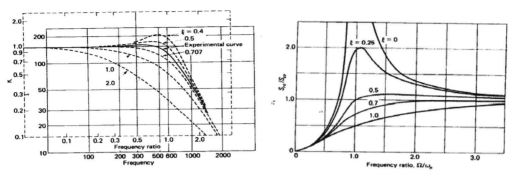

Figure 8.2. Plots of equations 8.5 and 8.6

2.1. VIBRATION VELOCITY SENSORS (SEISMOMETERS)

When the natural frequency of the sensor ω_n is low compared with the vibration frequency ω being measured, the ratio ω/ω_n is large. The amplitude of the relative displacement Z approaches that of the vibration A, regardless of the value of the damping ratio ζ. In these circumstances the mass m remains stationary while the surrounding case moves with the vibrating body. Sensors of this type are called seismometers. The relative motion z is usually converted to a voltage by making the seismic mass a permanent magnet, which moves relative to coils fixed to the case. Since the voltage generated is proportional to the rate at which the coils cut the magnetic field lines, the output of the sensor is proportional to the vibration velocity. A typical seismometer will have a resonance frequency between 1 and 5 Hz, and a useful bandwidth from 10 Hz to around 1 kHz.

The main disadvantage of seismometers as vibration sensors is their large size. Since $Z = A$, the relative motion of the seismic mass has to be of the same order as that of the vibration being measured. The sensor housing must be large enough to accommodate this motion.

2.2. ACCELEROMETERS

When ω_n is high compared to ω, the sensor output becomes proportional to acceleration. Examination of equation 8.5 shows that the factor $\sqrt{\left[1-(\omega/\omega_n)^2\right]^2+\left[2\zeta(\omega/\omega_n)\right]^2}$ approaches unity for $\omega/\omega_n \circledR 0$, so that

$$Z = \frac{\omega^2 A}{\omega_n^2} = \frac{acceleration}{\omega_n^2}.$$

(8.7)

Thus Z, the displacement amplitude of the seismic mass with respect to the case, is proportional to the acceleration of the motion being measured multiplied by a factor $1/\omega_n^2$. The useful range of the accelerometer is a plot of the following equation:

$$\frac{1}{\sqrt{\left[1-(\omega/\omega_n)^2\right]^2+\left[2\zeta(\omega/\omega_n)\right]^2}}.$$

(8.8)

For various values of damping ratio ζ, figure 8.2 shows that the useful range of frequencies for an undamped accelerometer is severely limited. However, when ζ reaches 0.7, the useful frequency range extends to around 20% of the resonance frequency, and within this range the maximum error is less than 0.01%.

2.2.1. Phase Distortion

The time delay between applying a mechanical input to an accelerometer and the appearance of the resulting electrical output is known as the phase shift. If this delay is not the same for all frequencies contained in the mechanical input, the phase relationship between the frequency components of the vibration waveform will be altered, and the resulting electrical output will become distorted. This effect is known as phase distortion, and to avoid it either the delay must be zero, or else all frequency components must be delayed by the same amount.

The first case, a zero delay or zero phase shift, corresponds to $\zeta = 0$ and $\omega/\omega_n < 1$ (see equation 8.6). However, as has been seen, zero damping is undesirable in an accelerometer. The second case, an equal time-wise phase shift applied to all frequency components, is almost satisfied when $\zeta = 0.7$ and $\omega/\omega_n < 1$. The phase angle ϕ is then given approximately by $\phi = \dfrac{\pi}{2}\dfrac{\omega}{\omega_n}$. Thus for $\zeta = 0.7$ (and $\zeta = 0$) phase distortion is almost eliminated.

2.2.2. Resonance Frequencies

The resonance frequency of an accelerometer is not constant, although a constant value will be specified on the sensor calibration chart. It depends not only on the seismic mass and the stiffness of the piezoelectric or other transducers to which it is attached, but also on the mass and stiffness of the vibration test object to which the device is attached, and to some extent on the stiffness of the mounting method used.

This situation is illustrated in figure 8.3. A seismic mass m_s rests on a transducer, such as a piezoelectric element, which is attached to the transducer base. K is the equivalent stiffness of the transducer(s) and its connection to the base. The mass of the base and housing is m_b. When the accelerometer is not coupled to any other object, as shown in figure 8.3a, the resonance frequency f_r is

$$f_r = f_s\sqrt{1+\dfrac{m_s}{m_b}}, \qquad (8.9)$$

where f_s is the resonance frequency of the seismic mass m_s on the stiffness K, which is given by

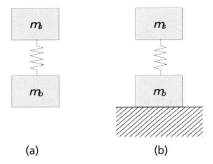

(a) (b)

Figure 8.3. Resonant frequency of an accelerometer

$$f_s = \frac{1}{2\pi}\sqrt{\frac{K}{m_s}}$$

(8.10)

From equation 8.9, it can be seen that the "free-hanging" resonance frequency depends upon the ratio of m_s to m_b. At first sight equation 8.9 seems to imply that the base of the accelerometer should be made as light as possible (m_b small), so that f_r is as large as possible and the usable bandwidth extended. However, when the accelerometer is mounted on a test object of large mass and stiffness as shown in figure 8.3b, m_b tends to infinity, and the accelerometer resonance approaches f_s. There is therefore no point in trying to make the base of the sensor light. Other considerations, such as the susceptibility to base strain, dictate the use of a stiff and consequently heavy base.

3. SENSING PRINCIPLES

By definition, an accelerometer requires a component, the seismic mass, whose movement lags behind that of the accelerometer's housing, which is coupled to the object under study. The ultimate goal of acceleration measurement is the determination of the mass displacement with respect to the housing. Any suitable displacement sensor capable of measuring microscopic movements under strong vibrations or linear acceleration can be used as an accelerometer. Differences in sensing principles give rise to variations in parameters such as bandwidth, sensitivity, or susceptibility to external interference. It is also important for the automotive engineer to be aware of all the features, both good and bad, of the many competing forms of accelerometer so that problems due to an inappropriate choice are avoided. Thus, some designs are well suited for some applications but will give poor performance in others. In this section the intention is to survey the types of accelerometer available and to highlight the good and bad features of each type of sensor.

3.1. PIEZOELECTRIC (PE) ACCELEROMETER

In 1880, the brothers Pierre and Jacques Curie discovered that charge separation occurs when pressure is applied to certain crystals, and that the same crystals change shape when a voltage is applied. This reciprocal effect has found many applications, including the crystal resonator in quartz watches. When pressure is applied to an asymmetrical crystalline material, it produces a charge separation or electric potential difference. The piezoelectric effect is the property of a crystal exhibited by the generation of

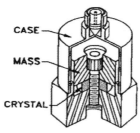

Figure 8.4. Schematic of piezoelectric accelerometer (*Source:* AIP Press Handbook of Modern Sensors: Physics, Designs and Applications, used with permission)

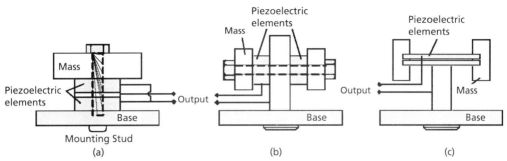

Figure 8.5. Design types for piezoelectric accelerometers

such a voltage when pressure is applied. Naturally occurring piezoelectric materials, such as quartz and Rochelle salt, produce small piezoelectric effects; whereas, synthetic piezoelectric compounds or ferroelectric ceramics, such as barium titanate and lead zirconate, have much higher responsivities and can be produced in any desired shapes and sizes. The sensitivities of these materials to stress is temperature dependent and disappears completely above a critical temperature called the Curie Point.

Accelerometers using the piezoelectric effect contain two elements: the mass on which the acceleration acts to produce a force, and a piezoelectric transducer that converts the force into electric charge. Three different design methods are used to stress the piezoelectric element. Common to all is the principle in which the accelerometer, when vibrated along its axis, causes the mass to exert a force on the piezoelectric element, which then develops a charge proportional to that applied force. These designs are based on compression, shear, and bending and are illustrated in figures 8.5a, b, and c, respectively. In figure 8.5a, the mass compresses piezoelectric elements mounted on a base, a technique that is low cost and has a high frequency range. In the shear technique (figure 8.5b), the elements are mounted between the masses and the base, which results in an accelerometer that is insensitive to base bending and has good thermal characteristics. The bending technique (figure 8.5c) has the elements isolated from the base between a split mass, which results in good low-frequency response and large output.

3.1.1. Frequency Response of Piezoelectric Accelerometers

The frequency response of a piezoelectric accelerometer is the frequency range over which the sensor provides a linear response. The upper end of the frequency response is limited by the mechanical stiffness and the size of the seismic mass, while the low-frequency limit is controlled by the amplifier roll-off and the discharge time constant. The greater the discharge time constant, the slower the signal leaks away and hence the better the low-end frequency response. The electrical output of a piezoelectric accelerometer as a function of frequency is typified by figure 8.6, which is obtained by applying a constant-amplitude sinusoidal acceleration and plotting the peak electrical output as a function of frequency. In practice, the useful bandwidth can extend from a few Hertz to about one third of the resonance frequency. Hence, piezoelectric accelerometers are used mainly for sensing frequencies in excess of 5–10 Hz. In automotive engineering, they are frequently used for sensing engine vibrations.

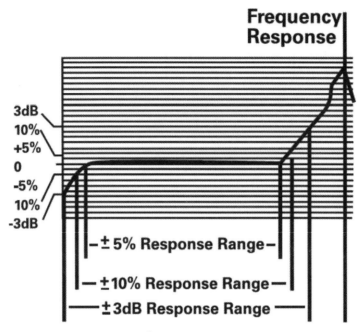

Figure 8.6. Accelerometer response versus frequency

3.1.2. Cross-axis Sensitivity of Piezoelectric Accelerometers

Most accelerometers are designed for motion sensitivity in one major axis. However, most accelerometers are also sensitive to shock and vibration inputs normal to the intended axis of sensitivity. Real accelerometers approach ideal behavior with varying degrees of accuracy. However, there is always some sensitivity to vibration components perpendicular to the intended sensing axis. This transverse or cross-axis sensitivity is minor and usually amounts to less than 5% of the major axis sensitivity. The dependence of reverse sensitivity on orientation is a consequence of the fact that the minimum charge and voltage sensitivity of the PE transducer elements is not always perfectly aligned with the axis of the accelerometer. This produces directions of maximum and minimum transverse sensitivity, and these are at 90° to each other and perpendicular to the main axis. If the axis of sensitivity is in any axis other than normal to the accelerometer base, it will be indicated by markings on the accelerometer case.

3.1.3. Micromachined Piezoelectric Accelerometer

A unique accelerometer fabricated by a combination of screen printing and silicon micromachining appears in figure 8.7 [4]. This new fabrication technology was first developed at the University of Southampton and used to drive a silicon micropump [5]. The device presented here is a deflection-based dynamic mechanical accelerometer with an inertial mass suspended by four beams, each located at a corner of the seismic mass. The deflection of the inertial mass relative to the chip frame due to applied accelerations is sensed using a thick-film printed piezoelectric element located on the supporting beams. This piezoelectric element is printed as a capacitor structure with the active layer being sandwiched between top and bottom electrodes. As the mass moves relative to the chip frame,

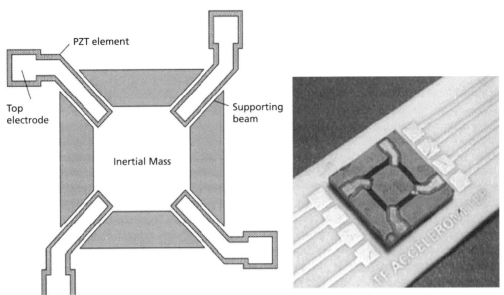

Figure 8.7. Screen printed and silicon micromachined piezoelectric accelerometer (*Source:* IOP, used with permission)

the beams deflect causing the piezoelectric layer to deform, hence, generating a charge. The amount of charge generated depends upon the piezoelectric properties of the printed layer and the deflection produced by the applied accelerations. The overall sensitivity of the accelerometer depends upon the magnitude of the charge generated in relation to the background noise associated with the sensor.

3.2. PIEZORESISTIVE (PR) ACCELEROMETERS

The piezoresistive effect occurs in silicon and other materials that change their resistances under physical pressure or mechanical work, so if a PR material is strained or deflected, its internal resistance will change and will remain changed until the original position is restored. In other words a change in electrical resistance occurs in response to changes in applied stress. Piezoresistive vibration sensors are formed by placing stress-sensitive resistors on highly stressed parts of a suitable mechanical structure. The PR transducers are usually attached to cantilevers or other beam configurations, and are connected into Wheatstone bridge circuits. The beam carries a seismic mass or may utilize its own self-weight so that under acceleration it deflects due to its inertia and undergoes structural changes. These stress variations are converted into an electrical output by the PR transducer. The signal is proportional to acceleration. As a sensing element a piezoresistive accelerometer incorporates strain gauges, which measure strain in mass-supporting springs. The strain can be directly correlated with the magnitude and rate of the mass displacement and subsequently with acceleration.

Figure 8.8 shows a wide dynamic range solid-state accelerometer developed by Endevco/Applied Signal Aerospace Company. When acceleration is applied along the sensitive axis, the inertial mass rotates around the hinge. The gauges on both sides of the hinge allow rotation of the mass to create compressive stress on one gauge and tensile on the other.

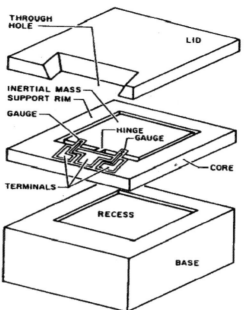

Figure 8.8. Exposed view of a piezoresistive accelerometer (*Source:* Endevco)

Piezoresistive accelerometers are relatively easy to construct, provide a frequency response down to DC, and work well over a relatively wide temperature range from about −50 °C to 150 °C. They find many applications in automotive systems such as active-suspension control, where the primary requirement is for good low-frequency performance. A further advantage making them valuable in automotive engineering is their ability to include signal processing and communication functions within the sensor package at little extra cost.

The drawbacks of PR devices are that the output signal level is moderate (typically 100 mV full-scale for a 10V bridge excitation), the sensitivity can be temperature dependent, which will adversely affect a PR accelerometer's repeatability, and the usable bandwidth is not as wide as that which may be obtained from a PE sensor.

3.2.1. Piezoresistors

If a rectilinear resistor has length L, width W, thickness t, and a bulk resistivity ρ, its resistance will be

$$R = \frac{\rho L}{Wt}. \tag{8.11}$$

The gauge factor or strain sensitivity is defined as k, where

$$k = \frac{dR/R}{\varepsilon}. \tag{8.12}$$

Here, ε is the relative change in the length of the resistor (the strain) due to a stress σ, applied to the substrate parallel to its length. Figure 8.9 shows the consequences of this applied stress. The length increases by an amount dl, while the width and thickness decrease by dw and dt according to the Poisson ratio v. Hence, $dw = -vw\varepsilon$ and $dt = -vt\varepsilon$. The original cross section was $A = wt$.

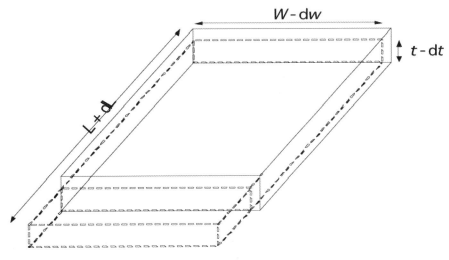

Figure 8.9. Consequence of applying stress to a piezoresistor

Referring to the strain ε, the new cross-sectional area is
$$A' = (W - dw)(t - dt) = wt + 2\nu wt\varepsilon + \nu^2 wt\varepsilon. \qquad (8.13)$$
However, the term $\nu^2 wt\varepsilon$ is very small compared with the other two terms in the equation and can be neglected, so that the change in the cross-sectional area becomes
$$A - A' = dA = -2\nu\varepsilon A. \qquad (8.14)$$
Rearranging equation 8.14 gives
$$\frac{dA}{A} = 2\nu\varepsilon, \qquad (8.15)$$
and differentiating equation 8.15 gives
$$\frac{dR}{R} = \frac{d\rho}{\rho} + \frac{dL}{L} - \frac{dA}{A}. \qquad (8.16)$$
Hence the gauge factor k is
$$k = \frac{d\rho/\rho}{\varepsilon} + (1 + 2\nu). \qquad (8.17)$$

Typically ν will be between 0.2 and 0.3. Equation 8.17 therefore shows that the longitudinal gauge factor is a function of changes in both longitudinal resistivity and geometry. In conventional foil or wire strain gauges piezoresistive effects are negligible and the variations in resistance are mainly a function of dimensional changes. For a foil gauge k is about 2. However, for piezoresistive strain gauges the first term in equation 8.17 is significant, and higher gauge factors can be achieved, giving enhanced sensitivity. It should also be noted that the resistivity of most PR materials is strongly temperature dependent, and that as a result PR strain gauges generally have higher thermal sensitivities that other types.

3.2.2. Silicon Piezoresistive Accelerometers

The first silicon accelerometer was demonstrated in 1979 [6]. It consisted of a single cantilever carrying PR strain gauges near its root. These devices were fragile and required the inclusion of a liquid-filled cell for damping. Improved designs have since appeared, and there are now three basic types of silicon PR accelerometer, as shown in figure 8.10. These are the single cantilever, the doubly supported structure, and the "top-hat" design. It will be noted that, despite its name, the single cantilever design (figure 8.10a) can have more than supporting beams carrying the seismic mass. The distinguishing feature of the single cantilever design is that the support beams are all placed along one edge. The doubly supported cantilever of figure 8.10b uses four supporting beams, and for this reason it is sometimes referred to as a quad cantilever. The top-hat approach is implemented when a device capable of undergoing large displacements is required. If the supports are folded as shown in figure 8.10c, the effective length can be increased while the package size is kept constant.

In all three types viscous damping is provided by the inclusion of a small volume of air. Among the three variations, the single cantilever has the highest gain, but can also have a large transverse sensitivity. The doubly supported and top-hat designs provide good off-axis cancellation and are reasonably robust. They are consequently the most common configuration for PR silicon accelerometers.

3.2.3. Resonance Frequency

The resonance frequency of a silicon PR accelerometer is determined by the stiffness of the support structure and the seismic mass as discussed in section 2. Typical resonance frequencies for silicon PR accelerometers are in the range of 500–5000 Hz. Hence, the bandwidth is considerably lower than that of a typical PE device. For comparison, thick-film devices usually have resonance frequencies between 300 and 2000 Hz.

3.2.4. Sensitivity

The sensitivity increases with seismic mass m_s, decreases with support stiffness K, and is modulated by a transduction efficiency term b, so the relationship can be represented as follows:

$$\text{Sensitivity} = b\frac{m_s}{K}. \qquad (8.18)$$

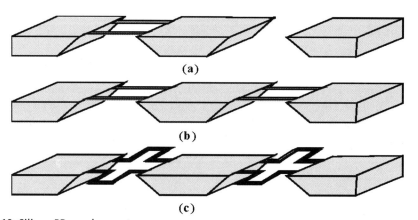

Figure 8.10. Silicon PR accelerometers

The main parameters determining b are the position and number of the PR transducers, and their transduction efficiencies, which are functions of their geometry and chemical constitution. The sensitivity is also inversely proportional to the square of the resonant frequency f_r:

$$\text{Sensitivity} = \frac{(2\pi)^2 b}{f_r^2}. \tag{8.19}$$

Equation 8.19 shows that if a sensitive accelerometer is required, the resonance frequency should be as low as possible. Thus the sensitivity requirement is in conflict with the need for a large bandwidth. A typical, doubly supported design with an output of 5 mVg^{-1} (millivolts per gravitational acceleration per volt of bridge excitation) will have a resonance of around 500 Hz, implying that the bandwidth extends from DC to around 150 Hz. If the design is modified so that the resonance moves to 37 kHz, the sensitivity decreases to about 5 μVg^{-1}.

3.2.5. Thick-film Piezoresistive Accelerometers

Thick-film circuits are formed through the deposition of layers of special pastes or inks onto insulating substrates. The printed pattern is fired in a manner akin to the production of pottery to produce electrical pathways having controlled resistances. Parts of a thick-film circuit can be made sensitive to strain or temperature. The thick-film pattern can also include mounting positions for the insertion of conventional silicon electronic devices, in which case the assembly is known as a thick-film hybrid. The process is relatively cheap, especially if large quantities are produced, and the use of hybrid construction allows the sensor housing to include sophisticated signal conditioning circuits. These factors indicate that thick-film technology is likely to play an increasingly important role in automotive sensor design [7].

Three main categories of thick-film inks exist: conductors, dielectrics (insulators), and resistors. Conductors are used for interconnections, such as the wiring of a bridge circuit. Dielectrics are used for coating conducting surfaces (such as steel) prior to laying down thick-film patterns; for constructing thick-film capacitors; and for insulating crossover points where one conducting path traverses over another. Resistor inks are the most interesting from the point of view of sensor design, since many thick-film materials are markedly piezoresistive.

The main constituents of thick-film inks are the binder (a glass frit), the vehicle (an organic solvent), and the active elements (metallic alloys or oxides). After printing, each layer of a thick-film pattern is dried to remove the vehicle, which gives the ink its viscosity. Drying also improves the adhesion properties, bonding the ink to its substrate and rendering the pattern immune to smudging. This stage is usually performed in a conventional oven at 100–150 °C.

A final high-temperature firing is required to remove any remaining solvent and to sinter the binder and the active elements. During the firing cycle, a thick-film pattern is raised to a temperature between 500 and 1000 °C. The glass frit melts, wets the substrate, and forms a continuous matrix, which holds the functional elements. The heating and cooling gradients, the peak temperature, and the dwell time determine the firing profile. This has a critical effect on the production of a thick-film circuit, since it allows the electrical characteristics of the inks to be modified. Resistor materials are especially sensitive to the firing profile, and the resistor layer is usually therefore the last to be fired.

However, the need for passivation of a circuit often necessitates covering it with a dielectric layer. To avoid changing the resistor values, a low-melting point dielectric is often used for the final layer. The need for high-temperature firing can cause problems if thick-film piezoresistors are to be applied to previously heat-treated components. The temperature used can adversely affect the properties of hardened or toughened steels, for example.

Thick-film circuits and sensors are created by screen printing [4]. This is essentially a stencil process, in which the printing ink is forced through the pattern areas of a mesh-reinforced screen onto the surface of a substrate. The screen stencils are formed by photolithography, in which a photosensitive mesh-filling material is exposed to ultraviolet light through a mask depicting the required pattern. The image is photographically developed, and those parts of the pattern that have not been fixed are subsequently washed away.

3.2.6. Thick-film Materials as Primary Sensors

The use of thick-film technology was introduced as a means of miniaturizing circuits without incurring the expense associated with fabrication in silicon. It was soon noted that thick-film materials had temperature- and stress-dependent properties. Although this was awkward from the point of view of circuit fabrication, it has since been turned to good account in sensor design. The linear temperature coefficient of resistance (TCR) possessed by certain platinum-containing conductive inks has allowed resistance thermometers to be constructed wholly in thick-film form [8]. More importantly, from the point of view of accelerometer design, the PR properties of thick-film resistor (TFR) inks can be used to form strain resistors. This approach has been used to make a number of pressure sensors [9], and by 1993 it was being exploited to produce accelerometers.

3.3. CAPACITIVE ACCELEROMETERS

Early accelerometers measured the displacement of the proof mass. Problems with this approach are the presence of a resonant peak, the difficulty of achieving good dynamic range, and the need to carefully calibrate the force versus displacement relationship. On the other hand, the powerful technique of force-balancing can reduce displacement of the proof mass to microscopic proportions. This is done by applying the output of a capacitive accelerometer to the input of a high-gain amplifier, the output of which feeds an electrical actuator that acts on the proof mass in the opposite direction to the initial excursion. This is a prime example of a negative feedback system, and if the actuator operates linearly, then its driving signal will be proportional to the force needed to minimize the initial excursion, from which the acceleration can be derived. The actuator must of course be capable of acting faster than the expected acceleration range, and a properly designed system will then allow the proof mass to remain almost stationary.

Capacitive displacement conversion is a proven and reliable method. A capacitive acceleration sensor essentially contains at least two components where the first is a "stationary" (i.e., connected to the housing) plate and the other is a plate attached to the inertial mass. These plates form a capacitor whose value is a function of a distance d between the two plates. Varying either the plate area (A) or the distance between the plates will linearly change the capacitor's value, and this can be accurately measured by an appropriate circuit:

$$C = \frac{\varepsilon_0 A}{d}. \tag{8.20}$$

Capacitive accelerometers are micromachined from single-crystal silicon, and the basic structure is typified in figure 8.11. Here, a conducting layer is deposited onto one surface of a silicon block, and a second conducting layer is laid down on one side of a second block, which acts as the seismic mass. This mass is supported by beams and is separated from the base by an air gap. The two halves of the

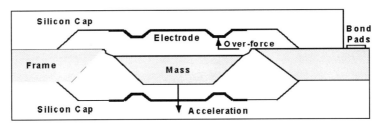

Figure 8.11. Capacitive sensing

sensor are electrostatically bonded together. Signal processing electronics can be incorporated within the sensor package if so desired.

The first commercially available, surface micromachined, complete acceleration measurement system on a single monolithic IC accelerometer was the Analog Devices ADXL50 [10] shown in figure 8.12. Surface micromachining builds layers of material on top of a silicon wafer and then selectively etches material away to make sensor structures, as opposed to machining material out of the wafer itself. It consists of a mass-spring system as well as a system to measure displacement and the appropriate signal conditioning circuitry. The mass is a bar of silicon, and the spring system is implemented by the four tethers that attach to each corner of the mass. The tethers are formed from crystalline silicon, a very stable and reliable spring element, and the strength of the tethers is sufficient to allow the device to withstand 200 g of physical shock. It responds to accelerations that occur in line with the length of the mass. When acceleration occurs, the mass moves with respect to the anchored ends of the tethers. The displacement of the mass is then proportional to the magnitude of the acceleration.

Figure 8.12 shows the parallel multiplate electrode structure of the Analog Devices surface-machined silicon accelerometer, the ADXL50, which has an overall size of 500 mm × 625 mm. Its

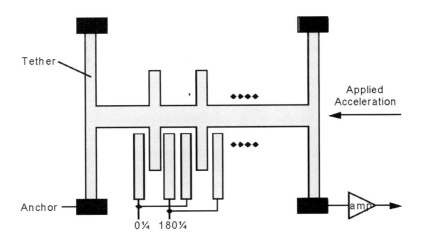

Figure 8.12. Analog Devices ADXL50 capacitive accelerometer (*Source:* "Capacitive Sensors" by L. Baxter. 1997. IEEE Series on Electronic Technology.)

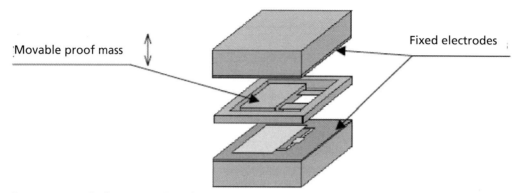

Figure 8.13. VTI bulk-micromachined capacitive accelerometer (*Source:* Nina HedbergVTI Technologies Oy, used with permission)

42 silicon fingers are 100 mm in length with a 2 mm gap and a total capacitance of 0.1 pF. The H-shaped piece is elastically mounted, taking advantage of the good spring characteristics of silicon, and responds to acceleration in the x direction with a small displacement. With a displacement in the $-x$ axis, the H piece picks up more of the 0° drive signal, and a demodulator (not shown) converts the displacement into acceleration. As the limiting resolution of the sense amplifier is 20×10^{-18} pF, a beam displacement of 20×10^{-12} m can be measured.

Capacitive accelerometers are expensive to manufacture and usually cost as much as the PE equivalent. The performance of these sensors is claimed to provide a frequency response down to DC, stable damping characteristics, and a useful bandwidth that is larger than PR but smaller than that provided by PE devices. In addition, their output is extremely stable over a wide temperature range. The output impedance of any capacitive sensor is intrinsically very high. However, if a capacitive accelerometer includes signal conditioning circuits within the package, a low final output impedance is usually provided so that vibration signals may be transmitted over lengthy cables without loss or distortion. To sum up, therefore, the inherent advantages of capacitive sensing are its low power, high output, wide dynamic range, and relative immunity to thermal effects. However, the signal processing requirements for this type of sensor are not straightforward, and it can be difficult to obtain a linear output.

3.3.1. Bulk-micromachined Capacitive Accelerometers

A low-cost, small and reliable automotive airbag accelerometer can be realized through bulk or surface micromachining techniques [11]. The capacitive sensing method is preferably chosen rather than the piezoresistive because of its inherent high sensitivity, low temperature shift, and capability of force-balancing and easy self-testing.

Bulk micromachining of single-crystal silicon results in a highly reliable mechanical sensor due to the stress-free and excellent mechanical properties of the material. High aspect-ratio structures are easily obtainable. Bulk micromachining refers to processes where the mechanical structures are etched out of an existing material. Typically, silicon or quartz are used as the bulk material and are valued for their near crystalline structure and excellent mechanical properties.

Another example of a bulk micromachined device is the VTI accelerometer shown in figure 8.13, and this also involves capacitive sensing. In this sensor, the body and the proof mass are insulated from each other and the capacitance is measured between them. As the distance between them decreases and increases, so the charge between them flows to and from the sensor, forming a measurable current that is proportional to the acceleration of the proof mass.

The acceleration-sensing element is made of single-crystal silicon and glass, which gives the sensor exceptional reliability, unprecedented accuracy, and outstanding stability in terms of time and temperature. The core of the VTI accelerometer is a symmetrical, bulk-micromachined acceleration-sensing element that has two sensing capacitors. Symmetry decreases temperature dependence and cross-axis sensitivity, and improves linearity. Hermeticity has been achieved by using anodic bonding to attach the wafers to each other. This facilitates the packaging of the element, improves reliability, and enables the use of gas damping in the sensor element.

3.4. INERTIAL ACCELERATION SWITCHES

In the mid-1990s, many automobile manufacturers implemented safety devices such as inertia switches that, in the event of a collision, shut off the fuel supply, unlocked the doors, and turned on the hazard lights. Sensing the rapid deceleration associated with a crash is crucial for correct operation of these switches. The sensors used must be able to discriminate between a crash and "legitimate" maneuvers such as driving over curbs, hitting potholes, or heavy braking.

Crash sensing simply requires a determination of when a preset deceleration threshold is exceeded, rather than an accurate measurement. The deceleration associated with a crash is at least an order of magnitude greater than anything likely to be experienced in normal driving. It is not considered economic to use accelerometers for crash sensing, since a fairly crude estimation of acceleration is sufficient. For this reason, inertia switches are often used instead.

One form of inertia switch, known as the "ball-and-tube" device, consists of a weight, often in the form of a steel ball or cylinder, which is fixed to one end of a spring [12]. The weight and spring are contained within a tube. When the assembly is accelerated or decelerated along its axis, inertial forces cause the weight to compress or extend the spring. The deflection of the weight is proportional to the acceleration and is sensed by placing an appropriate sensor in the side of the tube. An alternative

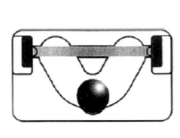 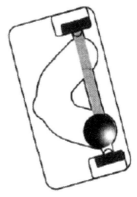

Figure 8.14. Ball-and-tube inertia device. V-shape gives 60° trigger angle; light beam is broken by ball; output can be electronically filtered to improve false triggering.

design, which has the advantage of being equally sensitive to any deceleration in a (normally horizontal) plane, uses a ferromagnetic sphere retained at the bottom of a conical or dish-shaped "saucer" by a permanent magnet. Rapid deceleration causes the ball to break free from the magnet, whereupon it rolls towards the rim of the saucer where it closes a switch.

Fuel-cutoff and battery-disconnection sensors operate in a similar fashion to airbag sensors, but are normally made more sensitive, with switching thresholds in the range of 8 g to 12 g.

3.5. THE OPTICAL ACCELEROMETER

Some applications cannot use piezoelectric or piezoresistive accelerometers because of their conductive construction and elements, their electrical sensing and signal transmission method, and possible interference with machine operation and safety hazards. In such cases, optical accelerometers provide a good alternative. The immunity of (mainly nonmetallic) optical accelerometers to electromagnetic interference [13] together with their ability to have the photo-emitter and the detector far from the accelerometer, such as when connected via optical fibers, make them extremely useful in harsh environments, including where strong electromagnetic fields are present. They are also useful where traditional sensors based on piezoelectric or capacitive working principles are not efficient. Optical accelerometers also have the advantage of being extremely lightweight, weighing as little as a fraction of a gram. Optical accelerometers can measure changes from 0 Hz up to resonant frequencies, and their minimal reliance on mechanical moving parts makes them very reliable and stable.

The operating principles of an optical accelerometer usually involve a light source with input and output guides to direct the light on to a sensing membrane, or a guiding path that changes its dimension or shape depending on the sensed acceleration. Though most optical accelerometer work is for

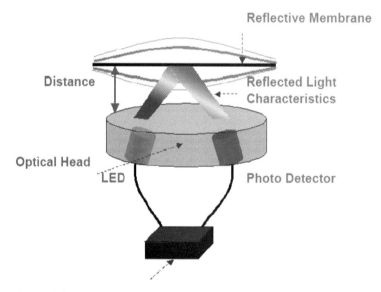

Figure 8.15. Light modulation in the optical accelerometer (*Source* Phone-Or, http://www.phone-or.com/temp/data/OA_WP.pdf, used with permission)

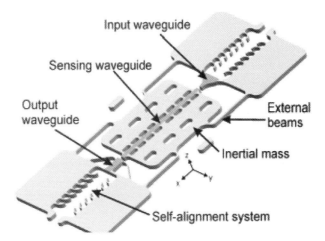

Figure 8.16. Polymer accelerometer with quad-beam structure and three-fold segmented waveguide. Sensing waveguide in the middle of the mass has been highlighted for clarification.

research purposes, there are nevertheless several good commercial products available. For example, the optical accelerometer by Phone-Or [14] uses an optical sensor that measures distance to a membrane by determining the intensity of light reflected from that membrane (see figure 8.15). Light emitted by a LED is beamed onto a highly sensitive membrane loaded with a tiny mass at its center in order to make it respond to vibration and acceleration. Because the membrane itself is able to sense even very weak transformations, the loaded mass weighs only a fraction of a microgram. Acceleration of the loaded membrane causes it to deflect very slightly, thereby changing the intensity of the light being reflected from its surface. This reflected light is transmitted to a photodetector, which uses simple electronic processing to convert the modulated light into an output signal.

The correlation between the changes in membrane position and the output signal is determined by the triangle formed by LED, membrane, and photodetector elements. As the accelerating membrane changes its position, the geometrical characteristics of the triangle also change, along with the light intensity reflected onto the photodetector. Given a predefined range of triangle coordinates, all membrane linear movements will result in corresponding linear changes to the output signal.

A unique polymer optical accelerometer described by Llobera and others, and shown in figure 8.16, consists of a three-fold segmented waveguide [13]. The input-output waveguides are located at the frame, while the center (sensing) waveguide is placed at the middle of the seismic mass. Since the whole structure has the same refractive index, some structure patterning is required to confine the light on the waveguide and not in the seismic mass. This has been accomplished by defining a periodicity of air gaps at both sides of the waveguide. This configuration assures that light mostly propagates through the waveguide, and that only a very small fraction of the total light is transferred to the mass (this light transference occurs at the region between two air gaps). When there is no applied acceleration, the waveguides are aligned and minimal losses can be expected. Under an applied acceleration, a misalignment between the waveguides is produced, causing an increase of the total losses in the device. Experimental results show that the device has a very high sensitivity, much higher than any previously reported optical accelerometer, with sensitivity between 6 and 11.5 dB/g, depending on whether positive or negative acceleration was applied.

4. SELECTION CRITERIA FOR ACCELEROMETERS

In order to perform adequately, each sensor must obviously be capable of accurately measuring the vibration frequencies and amplitudes of interest, and also be appropriate for use with instruments based on frequency range, size, weight, mounting, and cost.

To ensure that a selected accelerometer will serve its intended purpose, it is essential to understand the operating principles of the various types described in the previous sections. This is the first step toward making the best selection for the prospective application. The next step is to narrow the field of choice by evaluating criteria specific to the intended application, such as the sensor's sensitivity, amplitude, frequency and temperature ranges, and its physical characteristics, along with mounting options and the environmental conditions at the installation site. To assist this procedure, the following recommendations are offered, the majority of which come from the manufacturers' technical resources, including Dytran [15], Endevco [16,17], and Metra Mess-und Frequenztechnik Radebeul [18].

4.1. ENVIRONMENTAL CONDITIONS

Accelerometers are frequently used to make measurements under severe environmental conditions. In automotive applications these can include both high and low temperatures, typically ranging from –40 °C to +150 °C; severe shock loadings up to several hundred g; a wide humidity range (up to 100% RH); exposure to electromagnetic radiation and to potentially damaging chemicals such as gas, oil, hydraulic fluid, and water. It is therefore important that all accelerometers used in automotive applications are sealed, are as rugged as possible, and have a sensitivity to environmental conditions that is as low as possible.

4.1.1. Thermal Sensitivity

Any sensor must obviously survive the expected temperature extremes of the proposed application environment. However, changes in sensitivity can occur at different operating temperatures, and any such variation must be acceptable within the context of the measurement requirements. Most accelerometers will behave satisfactorily within the typical automotive temperature range of –40 °C to +150 °C. Adhering to the manufacturer's recommended operating temperature range will not only ensure accurate measurements, but also prolong the useful life of the sensor. For example, a quartz piezoelectric accelerometer should be operated in a constant temperature environment. Although quartz itself produces no output change with varying temperatures, the metallic preload acting on the quartz element will expand and contract with temperature fluctuations.

Accelerometers can also exhibit a slowly varying output when subjected to temperature shocks or transients. This output arises from two causes: a pyroelectric effect and nonuniform thermal expansion of the accelerometer structure, which subjects the sensing element to stress variation. The effect of a temperature transient is usually seen as a low-frequency electrical output, or drift, which can be removed by appropriate filtering. Where low-frequency measurements are made this effect must be considered more seriously, but fortunately this is not usually the case in automotive engineering.

4.1.2. Humidity

Most commercial accelerometers are of sealed construction, but may not be truly hermetic; that is, they are sealed with epoxy or other elastomers, or sometimes O-rings, but are not fusion sealed. This treatment may be adequate for most normal environments, but true hermetic sealing may be required when the sensor is to be used where wide temperature and humidity excursions are to be encountered. If it is necessary to immerse an accelerometer in a liquid, or to use it in an environment where heavy condensation is likely, it may be necessary to take special precautions with cable connectors and jackets, which must be able to withstand high humidity or wet environments. Fortunately, these are commercially available.

4.1.3. Acoustic Sensitivity

Accelerometers are somewhat sensitive to acoustic excitation. High-level acoustic noise impinging on the case of an accelerometer can cause it to generate an output similar to that of a microphone. Most manufacturers do not test their product designs for acoustic noise sensitivity because in most applications acoustic noise output is not a significant problem. In applications where the acoustic environment is greater than 100 dB SPL (sound pressure level), the recommendation is to use instruments that have been tested and demonstrated to have low acoustic sensitivity. If these data are not available, it is best to specify units with case strain sensitivity as low as possible. In applications involving 120 dB SPL, units tested for acoustic sensitivity should always be used.

4.1.4. Base Strain Sensitivity

When an accelerometer is mounted on the surface of a structure that is undergoing strain variations, some of the strain is transmitted into the accelerometer housing. Strain output may be in phase or out of phase with the acceleration output. Since the strain is most likely caused by structural vibration, the strain output will often be at the same frequency as the acceleration output. This calls for accelerometers with low base strain sensitivity, and this is usually specified by the manufacturer for individual accelerometers in units of acceleration g (9.8 m/s^2) per microstrain (μ).

4.1.5. Electromagnetic Interference

Strong magnetic fields often occur around spark ignition systems, electrical machines, and other electrical wiring at mains frequency (usually 50 or 60 Hz and multiples thereof). Most commercially available accelerometers are housed within welded steel or stainless steel enclosures. These normally provide adequate screening from the effect of electric or magnetic fields if shielded and grounded in accordance with the manufacturer's instructions. Stray signal pickup can be avoided by proper cable shielding, and this is of particular importance for sensors with charge outputs.

4.2. MOUNTING TECHNIQUES

Transducer mounting technique and surface preparation can affect the amplitude-versus-frequency response of the accelerometer, particularly at high frequencies [16]. The ability to couple motion, in the form of vibration or shock, to the accelerometer with high fidelity, is very dependent upon the method of mounting the instrument to the test surface. For best accuracy, it is important that the accelerometer mounting be tightly coupled to this surface to ensure the duplication of motion, especially at higher frequencies. Care should be taken to ensure a flush mating with a smooth, flat surface. Nicks, scratches, or other deformations of the mounting surface or the accelerometer do affect frequency response [15]. Since some mounting methods may adversely affect accuracy, it is important to understand the mechanics of these methods for best results— a poor mount can have an adverse effect on frequency response. Figure 8.17a schematically illustrates a condition where the accelerometer has acquired a "dished" shape through heavy usage. The mechanical analogy of this is a leaf spring with spring rate K_m as shown in figure 8.17b. There are now two spring-mass systems with this type of anomaly and both will affect frequency response. The new spring-mass system is formed by spring K_m and the mass of the entire accelerometer M_m. The resonant frequency f_m of this new system will most likely be lower than that of the accelerometer and may affect the response curve as illustrated in figure 8.17c. Even though the new resonant frequency is higher than the actual resonance of the accelerometer, its effect will be to increase the output at the high frequency end of the accelerometer response.

4.2.1. Stud Mounting

The commonly recommended and preferred method of mounting an accelerometer uses a steel stud. The stud may be integral, that is, machined as part of the accelerometer; or it may be separate (removable) as shown in figure 8.18. This mounting technique provides higher transmissibility than other methods. The transducer should be mounted with appropriate adapter studs or bushings, so ensuring that the entire base of the transducer has good contact with the surface of the test surface. The inclusion of a thin layer of silicone grease between mating surfaces improves the mounting stiffness and hence the high-frequency performance.

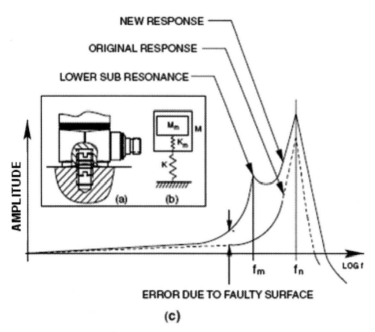

Figure 8.17. Frequency response of acceleration measurement (*Source:* Dytran Instruments, Inc., used with permission)

ACCELEROMETERS • 225

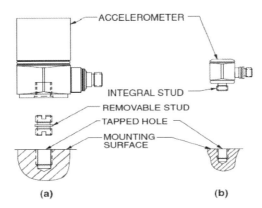

Figure 8.18. Methods of stud mounting (*Source:* Dytran Instruments, Inc., used with permission)

4.2.2. Adhesive Mounting

Most miniature accelerometers can be mounted using only an adhesive, or situations may arise where the stud mount method is impractical or even impossible, such as when mounting the accelerometer to thin sheet metal or to surfaces where drilling a mounting hole is not allowable. In such cases, an adhesive mount installation may be the best, and sometimes the only, method available. Four typical methods of mounting accelerometers are depicted in figure 8.19.

The natural frequency of a mounted accelerometer is dependent on the stiffness of the mounting. Given that no adhesive is as stiff as a mounting stud, it must be accepted that the more adhesive joints between the test structure and the transducer, the more degradation in motion transmissibility will occur. Adhesives for mounting accelerometers include cyanoacrylate instant adhesive, magnets, double-sided tape, petro-wax, and hot glue [17].

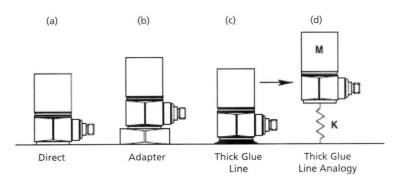

Figure 8.19. The adhesive mount, direct and with adapter (*Source:* Dytran Instruments, Inc., used with permission)

4.2.3. Magnetic Mounting

Magnetic mounting adapters, such as those seen in figure 8.20, are popular for applications requiring periodic quick point-to-point measurements. Most magnetic adapters are massive and useful only for measurements below a few hundred Hertz. In general, magnetic adapters should be used with caution and rarely trusted at frequencies above 1 kHz. A response degradation in direct proportion to the weight of the accelerometer must be expected.

4.3. INTERCONNECTING CABLES

Piezoelectric and capacitive sensors are very high impedance devices. They are therefore susceptible to triboelectric noise, which can be generated by connecting cables subjected to mechanical motion. Dynamic bending, compression, or tension of cables momentarily separates the cable screen from the dielectric, and this results in local capacitance changes and the release of triboelectric charge. This problem is especially severe at low frequencies, especially if conventional coaxial cable is used rather than specially treated accelerometer leads. These cables should be clamped to the vibrating specimen using epoxy, glue, wax, or adhesive tape so that relative movement is avoided as far as possible. Various methods of securing the cables are shown in figure 8.21.

By nature, cables are usually the weakest part of any measurement system, and because they are somewhat fragile, they must be handled with care. The weakest point on the cable assembly is the intersection of the wire and the cable connector, so excess stress at this point must be minimized. Sharp

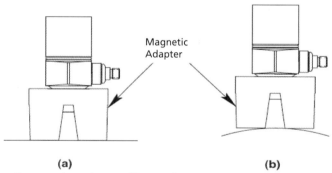

Figure 8.20. Magnetic mounting adapters (*Source:* Dytran Instruments, Inc., used with permission)

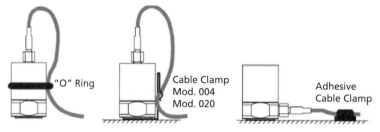

Figure 8.21. Methods of cable tie-down (*Source:* Metra Mess-und Frequenztechnik Radebeul, used with permission)

bends here can overstress conductors and insulation, causing cable failure. Constructional material for cables should be selected to insure survivability in the environments in which they are used. Teflon cables are suitable for high-temperature use or where exposure to harsh chemicals is of concern. Sealed, integral cables should be used for sensors that are submerged or installed in areas of high humidity or subjected to wash-down.

4.4. PREAMPLIFIERS

The preamplifier itself may take the form of either a voltage or a charge amplifier. As the names suggest, the former accepts an input voltage signal and the latter an input charge. Both produce a voltage output at a low impedance level. However, the effect of input cable length, and hence capacitance, is negligible for the charge amplifier, whereas the voltage amplifier is markedly sensitive to input capacitance, which limits the associated cable length. Furthermore, the input resistance of the voltage amplifier may not be sufficiently high to be neglected, and may result in a poor low-frequency response. Hence, charge amplifiers are preferred because the charge remains invariant almost irrespective of cable length, though it is unwise to use a cable longer than about 10 m [18]. Nevertheless, because voltage amplifiers contain fewer components, they are often less expensive and more reliable and so may be specified for some applications.

Piezoelectric transducers generate charge output signals generally around a picoCoulomb (1 pC or 1000 fC) at a very high impedance. Consequently, a charge preamplifier must be used that converts

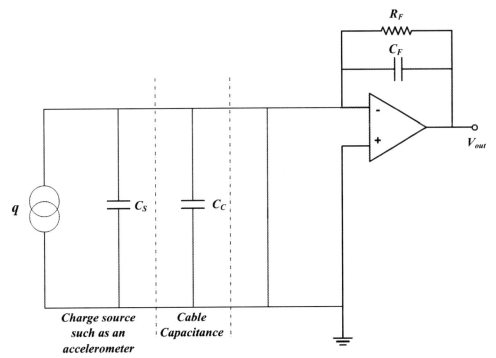

Figure 8.22. Charge amplifier

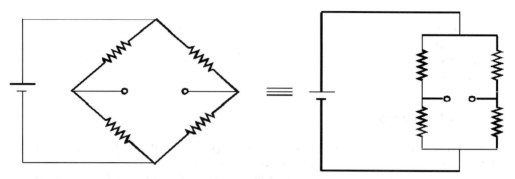

Figure 8.23. Wheatstone bridge

an input charge to an output voltage. A relevant circuit is shown in figure 8.22, where the effective input capacitance $C_p = C_F(G-1)$, where G is the open loop gain of the operational amplifier involved. (Here, the "real" input capacitance of the amplifier has been neglected because it is very small.) The output voltage of the system is given by

$$V_{out} = \frac{qG}{C_S + C_C - C_F(G-1)}. \qquad (8.21)$$

The cable capacitance C_C, source capacitance C_S, and op-amp input capacitance C_p are usually much smaller than C_F. Since the open-loop gain of an op-amp is very high, the output becomes

$$V_{out} = \frac{q}{C_F}. \qquad (8.22)$$

With C_F constant, the output voltage V_{out} is proportional to the input charge q. Only when the cable becomes so long that the value of C_C approaches that of C_F will the sensitivity of the circuit be affected.

The piezoresistive sensor is usually connected as part of a Wheatstone bridge circuit containing one, two, or four active sensors, and which can provide first-order temperature compensation. A bridge circuit is essentially a pair of parallel potential dividers, as shown in figure 8.23, and an excitation voltage must be applied before any output signal is obtained. Changes in the sensor resistance produce changes in the out-of-balance voltage that can be amplified by a conventional high-impedance voltage amplifier, usually an operational amplifier.

4.5. CHOOSING THE CORRECT ACCELEROMETER: A SUMMARY

Proper sensor selection necessitates the careful evaluation of the desired measurement required by the proposed application. Data estimations based on dynamic measurement considerations and environmental concerns can be used as starting references for specifying the most appropriate accelerometer. For example, minimum speeds and any subharmonics relate to the low-frequency response of the accelerometer, while the highest harmonic of the measurement relates to the high-frequency response [19]. In choosing the proper sensor, the following points can be used as a guide:

- The accelerometer mass must be as small as possible in comparison with the mass of the structure to be measured.
- The dynamic amplitude range must be able to accommodate the maximum acceleration level anticipated.
- The frequency response must match the overall frequency spectrum to be measured. Accelerometers should be used only for the measurement of frequencies specified by the manufacturer; each accelerometer has a usable frequency band over which its response is optimized. Below and above this band correct operation cannot be guaranteed. Also, in the region above the usable frequency, the internal resonant frequency of the instrument may result in an incorrect, usually high, response at such frequencies.
- Environmental factors such as high temperatures can have a deleterious effect on the measurements.
- Cross-axis sensitivity of the accelerometer is a measurement of how well the accelerometer measures vibration along the intended axis and does not pick up vibration in other axes. Ideally, cross-axis sensitivity should be zero, but in reality most accelerometers have cross-axis sensitivities of about 3%–4% of those along the intended axis. Anything in excess of 10% should be avoided.

REFERENCES

1. Weinberg, H. 2002. MEMS sensors are driving the automotive industry. http://www.sensorsmag.com/articles/0202/36.
2. Yazdi, N., F. Ayazi, and K. Najafi.1998. Micromachined inertia sensors. *Proceedings of the IEEE* 86(8): 1640–59.
3. World acceleration, vibration, and velocity (AVV) market research report. http://www.researchandmarkets.com/reports/365386.
4. Beeby, S. P., J. N. Ross, and N. M. White. 2000. Design and fabrication of a micromachined silicon accelerometer with thick-film printed PZT sensors. *Journal of Micromechanics and Microengineering* 10: 322–28.
5. Koch, M., N. Harris, R. Maas, A. G. R. Evans, N. M. White, and A. Brunnschweiler. 1997. A novel micropump design with thick-film piezoelectric actuation. *Measurement Science and Technology* 8: 49–57.
6. Roylance, L. M., and J. B. Angell. 1979. A batch fabricated silicon accelerometer. *IEEE Transactions on Electron Devices* 26: 1911.
7. White, N. M., and J. D. Turner. 1997. Thick-film sensors: Past, present and future. *Measurement Science and Technology* 8: 1–20.
8. Reynolds, Q. M., and M. G. Norton. 1985. Thick film platinum temperature sensors. *Proceedings of Test and Transducer Conference* 2: 31–44.
9. Prudenziati, M., ed. 1994. *Thick film sensors*. Amsterdam: Elsevier.
10. Analog Devices. 1996. ADXL05 Datasheet. http://www.analog.com/en/index.html.
11. Tsugai, M., Y. Hirata, K. Tanimoto, T. Usami, T. Araki, and H. Otani. Airbag accelerometer with a simple switched-capacitor readout ASIC. 1997 *SPIE* 3224: 74–81.
12. Turner, J. D., and L. Austin. 2000. Sensors for automotive telematics. *Measurement Science and Technology* 11: R58–79.
13. Llobera, A., J. A. Plaza, I. Salinas, J. Berganzo, J. García, J. Esteve, and C. Domínguez. 2004. Characterization and passivation effects of an optical accelerometer based on antiresonant waveguides. *IEEE Photonics Technology Letters* 16(1): 233–35.
14. Phone-Or Ltd. An optical revolution for the accelerometer. White paper. http://www.phone-or.com.
15. Dytran Instruments, Inc. Accelerometer mounting considerations. Technical paper 2008. http://www.dytran.com/go.cfm/en-us/content/tech-education-a8/x?SID=.

16. Endevco Corporation. Guide to accelerometer installation 2008. Technical paper 319. http://www.endevco.com/resources/tp_pdf/TP319.pdf.
17. Endevco Corporation. Guide to adhesively mounting accelerometers 2008. Technical paper. http://www.endevco.com/resources/tp_pdf/TP312.pdf.
18. Metra Mess-und Frequenztechnik Radebeul, AN9E application note piezoelectric accelerometers accelerometer cabling. http://www.mmf.de/PDF/AN9E-Accelerometer_Cabling.pdf.
19. Eric S. Industrial Monitoring Instrumentation (IMI) Accelerometer Selection Criteria. http://www.davidson.com.au/products/vibration/pcb/imi/theory/selecting-installing.asp.
20. Turner, J. D. 1988. *Instrumentation for engineers*. London: Macmillan.
21. Fraden, J. 1997. *Handbook of modern sensors: Physics, designs and applications*. 2nd ed. New York: American Institute of Physics Press.
22. Norton, H. N. 1982. *Sensor and analyzer handbook*. Englewood Cliffs, NJ: Prentice Hall.
23. Soloman, S. 1999. *SENSORS handbook*. New York: McGraw-Hill.
24. Seippel, R. G. 1983. *Transducer, sensors and detectors*. Reston, VA: Reston Publishing Company.
25. Baxter, L. K. 1997. *Capacitive sensors: Design and applications*. Piscataway, NJ: IEEE Press.

CHAPTER 9

GAS COMPOSITION SENSORS

Jonathan Swingler

1. INTRODUCTION

The use of gas sensors in automotive applications is increasing. For many years the main application area has been for monitoring exhaust gases formed from the combustion of gasoline or diesel fuel [1]. Recently, however, gas sensors have been applied to other systems such as climate control within the passenger cabin. As the motor vehicle becomes more sophisticated, further new applications are likely to appear.

On-board gas sensors for monitoring exhaust gases are essential in a modern vehicle. They ensure that the engine runs at optimum performance and that emissions of particular pollutants are kept to a minimum. In addition, a selection of sensors is required for use when the vehicle undergoes testing during servicing and repairs.

In this chapter an introduction is given to the basic principles of power production by the combustion of fuels, pollution problems that arise, and legislation that has been adopted to minimize these pollutants. Secondly, the main pollutants are discussed, including such gases as carbon dioxide (CO_2), nitrogen oxide compounds (NO_x), carbon monoxide (CO), and hydrocarbon compounds (HC). A discussion of particulates is also given, and sensor technologies used to identify and quantify these substances are highlighted. Thirdly, the most important sensor technology in the automotive application, the exhaust gas oxygen (EGO) sensor, is described in detail, including its principle of operation in the vehicle. The chapter concludes with some perspectives on novel technologies and applications in the motor vehicle.

2. POWER PRODUCTION AND POLLUTION PROBLEMS

The first gas sensors were incorporated into the management of power production and pollution control in the mid-1970s. Before that date few electronic systems were used on motor vehicles. The only electronic devices used were alternator diodes and voltage regulators [1,2]. However, the advent of the gas sensor and its associated electronics has allowed much better control of both gasoline and diesel internal combustion engines.

2.1. GASOLINE AND DIESEL ENGINES

The internal combustion (IC) engine is the most commonly used power source for the motor vehicle today. Power is generated by converting chemical energy stored in a liquid fuel to mechanical work by combustion. Table 9.1 classifies engines for both internal and external combustion. Internal combustion is an open process where the working medium (i.e., the combustion gas) is not reused but is lost to the environment. This working medium undergoes chemical change by combustion to produce work energy.

Table 9.1. Type and classifications of engines

	Internal combustion				External combustion	
Type of process	Open process where combustion gases are the working medium and are chemically changed in the process				Closed process where combustion of a second medium heats a working medium which remains chemically unchanged	
					No phase change	Phase change
Type of combustion	Cyclic combustion				Continuous combustion	
Type of ignition	Auto ignition	External ignition				
Type of machine — Cylinder	Diesel	Hybrid	Otto	Rohs	Stirling	Steam
Type of machine — Turbine	—	—	—	Gas	Hot steam	Steam
Type of mixture	Heterogeneous			Homogeneous	Heterogeneous	

2.1.1. Gasoline or Petrol fuels

Gasoline fuels contain many hydrocarbon types with molecule lengths from 3 to 12 carbon atoms. Boiling points can range between 25 °C and 215 °C. Gasoline is classified by its antiknock quality, represented by the RON (Research Octane Number), or MON [3]. The two main types of unleaded gasoline specified by European Standard EN228 are

        Regular   minimum RON91   42.7 MJ/kg
        Premium   minimum RON95   43.5 MJ/kg

Fuels with higher RONs have better antiknock performances. These RONs are determined empirically, and a gasoline with an octane number of 95 has the performance of an untreated gasoline with a volumetric content of 95% iso-octane (trimethyl pentane) in a mixture of iso-octane and n-heptane.

2.1.2. Diesel Fuels

Diesel fuels contain longer-chain hydrocarbons than gasoline with boiling points between 180 °C and 370 °C. Because the diesel engine does not use spark ignition, the fuel must ignite spontaneously when compressed within the cylinder. The Cetane number (CN) is a measure of the ability of the fuel to self-ignite. Diesel fuels with higher CNs have better self-ignition properties. A Cetane number of 100 is given to n-hexadecane (cetane), which ignites very easily, whereas methylnaphthalene is given a Cetane number of 0 because it is very difficult to self-ignite. A typical diesel fuel has a CN of 49 giving 42.5 MJ/kg.

2.1.3. Additives to Fuels

Additives are included into fuel for many reasons, such as providing antiknock and antiaging qualities, detergents, corrosion inhibitors, and icing protection. A typical fuel might have the following additives:

- Antioxidants (oxidation inhibitors)
- Corrosion inhibitors (such as carboxylic acids and carboxylates)
- Metal deactivators (chelating agents)
- Demulsifiers (such as polyglycol derivatives)
- Antiknock compounds (such as lead alkyls)
- Anti-icing additives (such as surfactants, alcohols, and glycols)
- Dyes
- Markers
- Drag reducers

2.1.4. Alternative Fuels

Ethanol and methanol have been proposed as alternative fuels since they reduce engine emissions, particularly CO_2. The basic idea is that vegetable matter should be grown specifically to produce fuel. As it grows, the crop removes CO_2 from the atmosphere during photosynthesis. When harvested, the crop is converted into alcohol fuel, which, when burnt, returns CO_2 to the atmosphere with no net increase in the amount of CO_2.

However, there are several problems with using these alcohol fuels in their pure form. Firstly, from an economic point of view, they are significantly more expensive to produce than gasoline or diesel fuels. Secondly, from a technological point of view, they have several performance problems. There is difficulty with cold starting and air mixing due to high latent heat of evaporation. Thirdly, the combustion of such fuels requires about twice the volume for the same power output compared with gasoline. One solution is to mix them with gasoline, and M85 is a fuel with 85% methanol and 15% gasoline [4].

2.1.5. Chemical Equations of Combustion

The main constituents of gasoline are heptane and octane, with the ratio dependent upon the performance required and the manufacturer. These constituents fully combust as follows:

$$C_8H_{18} + 12\tfrac{1}{2}\, O_2 \rightarrow 9\, H_2O + 8\, CO_2 \quad 44.6\ \text{MJ/kg}$$

$$C_7H_{16} + 11\, O_2 \rightarrow 8\, H_2O + 7\, CO_2 \quad 44.4\ \text{MJ/kg}$$

The main constituents of diesel fuel are hexadecane and methyl-naphthalene, with the ratio again dependent upon the performance required and the manufacturer. These constituents fully combust as follows:

$$CH_3(CH_2)_{14}CH_3 + 24\tfrac{1}{2}\, O_2 \rightarrow 17\, H_2O + 16\, CO_2$$

$$C_{11}H_{10} + 13\tfrac{1}{2}\, O_2 \rightarrow 5\, H_2O + 11\, CO_2$$

These can be generalized in the following form [1]:

$$C_mH_n + (m + n/2)\, O_2 \rightarrow m\, CO_2 + n/2\, H_2O$$

If the characteristics of a fuel are known, the exact amount of oxygen needed for full combustion can be determined. The sole products of full combustion are carbon dioxide and water vapor. In an ideal combustion process there are no other emissions.

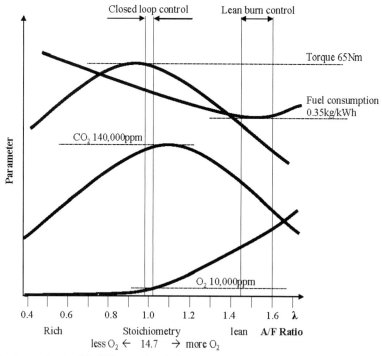

Figure 9.1. Engine output with different Lambda

2.1.6. The Normalized Air-fuel Ratio Lambda (λ)

A typical gasoline engine is expected to use 14.7 kg of air to fully combust 1 kg of fuel (an air-fuel ratio of 14.7/1). This ideal scenario is known as the stoichiometric point in the air-fuel ratio, when all the oxygen in the air mixed with the fuel is used. With this information, an engine can be operated for optimum performance. An air-fuel ratio with a higher fuel amount compared to the stoichiometric point is known as a "rich" mixture, whereas an air-fuel ratio with a lower fuel amount compared to the stoichiometric point is known as a "lean" mixture.

The "normalized air-fuel ratio," Lambda (λ), is defined in equation 9.1:

$$\lambda = \frac{actual\ air\ /\ fuel\ ratio}{stoichiometric\ air\ /\ fuel\ ratio}. \tag{9.1}$$

This is the ratio of actual or current air-fuel ratio compared to the stoichiometric ratio of air to fuel. When the engine is running at the stoichiometric point, then $\lambda = 1$.

However, during combustion of the air-fuel mixture in the engine cylinder, not all the available oxygen is used, even in a rich fuel mixture. This is due to a number of factors, such as insufficient time the mixture is in the engine cylinder for complete combustion, various flow processes within the cylinder, and so forth. Also, if the mixture is very lean, not all the fuel will undergo combustion, again due to many factors.

At the stoichiometric point of 1 kg of fuel to 14.7 kg of air, it is typical for 0.5%–1% of O_2 to be found in the exhaust. Figure 9.1 shows how the amount of O_2 in the exhaust varies with the Lambda

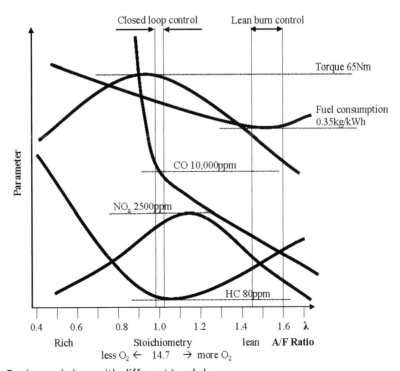

Figure 9.2. Engine emissions with different Lambda

value. Under increasing lean-burn conditions, when there is less fuel in the mixture (more O_2 available), increasing amounts of O_2 are detected in the exhaust.

Most engine management systems operate either a "closed-loop control" method around $\lambda = 1$ and/or a "lean-burn control" method at around $\lambda = 1.5$ (see section 5). As shown in figure 9.1, a typical gasoline engine produces maximum torque output when the air-fuel mixture is slightly rich, but maximum production of CO_2 appears when the air-fuel mixture is slightly lean. Best fuel consumption is found in the lean region around $\lambda = 1.5$. Some typical values for these parameters are included in figure 9.1.

Figure 9.2 shows the concentrations of three pollutants produced from a gasoline engine at different Lambda levels. The HC pollutants are at a minimum close to the stoichiometric point, whereas maximal amounts of the NO_x pollutants are produced at this point. The leaner at which the engine operates, the less CO pollutant is formed. Some typical values of these parameters are also given in the figure.

2.2. LOCAL POLLUTION

The atmosphere of earth consists of 21% oxygen (O_2), 78% nitrogen (N_2) and 1% water vapor (H_2O), plus trace gases, by volume. Trace gases in dry air are listed in table 9.2. Argon makes up nearly 1% by volume, and CO_2 is about 300 ppm depending on local conditions.

Table 9.2. Trace gases in Earth's atmosphere

TRACE GAS	CONCENTRATION IN DRY AIR (%)		CHEMICAL SYMBOL
	BY VOLUME	BY WEIGHT	
Argon	0.933	1.28	Ar
Carbon Dioxide	0.03	0.046	CO_2
Neon	0.0018	0.0012	Ne
Helium	0.0005	0.00007	He
Krypton	0.0001	0.0003	Kr
Hydrogen	0.00005	—	H_2
Xenon	9 10–6	0.00004	Xe

Motor vehicle pollutants that concentrate in a particular locality can lead to smog and acid rain, both with resultant health concerns. This is not a new problem—in London in the late 1800s, smog was a mixture of smoke and fog, which caused great health problems. This smog was caused by high emissions of sulfur dioxide (SO_2) and soot from burning coal. Soot promotes the condensation of water droplets in air forming fog, after which sulfur dioxide dissolves in the droplets to form sulfurous and sulfuric acids. This was the once-notorious London smog.

Smog can also be formed through photochemical processes in a fog-free atmosphere containing the NO_2, CO, and HC produced by the automobile. The well-known Los Angeles smog formed in the 1960s in localized areas of California due to high vehicle emissions and high levels of sunlight.

The typical pollutants from a motor vehicle are shown in table 9.3.

Table 9.3. Typical pollutants

Pollutant	Typical Exhaust
NO_x	2500 ppm
HC	80 ppm
CO	10000 ppm
O_3	0
Particulates	—

Nitrogen oxides (NO_x) are the main pollutant from gasoline engines and are formed during high-temperature combustion processes in air. They consist mainly of three gases, N_2O, NO, and NO_2. Emissions of NO_x in Europe reached a maximum in 1989 at 28 M tonnes per year [5].

Hydrocarbons (HC) emissions are compounds resulting from incomplete combustion, and in Europe they reached a peak in 1989 at 23 M tonnes per year being produced [5].

Ozone (O_3) is not an emission from the motor vehicle's engine but is formed from air-chemical processes triggered by radiation from the sun. The ozone-forming chemistry is not fully understood, but it is thought that vehicle emissions contribute to its formation. Ozone has advantages and disadvantages. The ozone layer in the upper atmosphere, the stratosphere, protects life from harmful UV radiation and so is beneficial. However, in the lower atmosphere (the troposphere) ozone is not so desirable. In excessive concentrations it can irritate the human respiratory tract and damage plants.

The size of particulate pollutants can range from nanometers to several millimeters. PM_{10}s are those of less than 10 micrometers in diameter and are of concern as they do not quickly deposit out of the atmosphere but remain airborne.

2.3. GLOBAL POLLUTION

Global pollution and changes in the atmospheric environment are important issues. The so-called Greenhouse effect is one such concern. A limited Greenhouse effect is required for the earth to maintain a particular temperature. Too high a concentration of Greenhouse gases and the earth will get too hot; too low and the earth may get too cold. It is thought that without any Greenhouse effect the average temperature of the surface of the earth would be about —18 °C, whereas we see an average of 15 °C. The main Greenhouse gases are water vapor and CO_2, but other trace gases also exhibit this Greenhouse behavior.

The most common Greenhouse gases produced by gasoline vehicles are listed in table 9.4 with estimates of their concentrations in the atmosphere. No data is available on water vapor. Concentrations of N_2O and O_3 in the troposphere (lower atmosphere) and CH_4 all contribute substantially more to the Greenhouse effect than does CO_2.

The concentration of CO_2 in the troposphere has been reported as increasing from 320 ppm in the 1960s to 360 ppm in the 1990s [6,7]. Resulting from all the activities on the planet, it is estimated that there is an increase in CO_2 in the troposphere of approximately 0.45% each year [8].

It has been calculated that around 3 Giga-tonnes per year of carbon have been added to the atmosphere throughout the 1980s and 1990s from human and natural processes. There are about 750 Giga-tonnes of carbon in the form of CO_2 in the troposphere. The oceans contain about 40 Tera-tonnes of carbon, and the biosphere holds about 1 Tera-tonne.

238 • AUTOMOTIVE SENSORS

Table 9.4. Summary of emission level for gasoline vehicles [1]

	Year	NO_x	HC	CO	Unit
California	1993	0.4	0.25	3.4	g/mile
10%	1994	0.2	0.125	3.4	g/mile
25%	1997	0.2	0.075	3.4	g/mile
2%	1997	0.2	0.04	1.7	g/mile
2%	1998	0.0	0.0	0.0	g/mile
USA	1983	1.0	0.41	3.4	g/mile
Japan	1994	0.4	0.25	3.4	g/mile
Europe	2003	0.2	0.125	1.7	g/mile
	1984	0.25	0.25	2.1	g/mile
	1992	0.15	0.97 NO_x	2.72	g/km
	1996	0.15	0.5 NO_x	2.2	g/km
	2000	0.08	0.12	2.3	g/km
	2000		0.20	2.3	g/km
	2005		0.10	1.0	g/km

2.4. LEGISLATION

The problem of local pollution by the motor vehicle has existed for many years. It was not until the 1960s, however, when the Los Angeles Bay area of California suffered from severe air pollution, that it was recognized that something had to be done. In 1966, the first legislation to restrict exhaust emissions from vehicles was adopted in California. CO and HC were initially controlled, but NO_x and O_3 soon followed.

In the 1970s, Californian emission control standards were adopted by the whole of the United States. These laws led to improvements in the control of fueling and ignition timing and the introduction of exhaust gas recirculation and catalysis to further reduce pollutants. Exhaust catalysts required unleaded fuels, since otherwise they become "poisoned" and made ineffective by lead deposits.

Europe also adopted emission controls in the 1970s, followed in the 1980s by the implementation of the "1973 US" noncatalyst standard in Switzerland and Sweden. At that time, lean-burn engines were investigated in Europe as a means to improve efficiency by up to 10% and the reduction of NO_xs. However, the adoption of further U.S. standards in Europe in 1993 led to abandonment of lean-burn engine development in favor of three-way catalyst engines. In these engines the O_2 content of the exhaust is monitored and the air-fuel mixture changed appropriately to give the correct control of CO, NO_x, and HC emissions.

Lean-burn technology still offers some advantages, however. At light loads the lean-burn systems perform well, and the Toyota lean-burn system [9] has been extended to include high-load conditions by changing over to a catalyst system.

3. EXHAUST EMISSIONS AND DETECTION DEVICES

In this section the main emissions from gasoline and diesel engine exhausts are highlighted and methods of gas detection discussed.

Carbon dioxide and water vapor are the main products of the combustion of fuels. In addition, nitrogen oxides, carbon monoxide, hydrocarbons, and particulates are formed. Each of these is discussed, and a table is given that shows whether the gasoline or diesel engine is particularly good or bad at producing these pollutants.

The exhaust gas oxygen (EGO) sensor is still the main transducer used in most engine management systems. However, in this section other emission sensors are discussed, most of which are still in the experimental stage of development for automotive applications.

3.1. CARBON DIOXIDE AND WATER VAPOR

Carbon dioxide (CO_2) and water vapor (H_2O) are both colorless and odorless compounds. As water vapor leaves the exhaust system and drops in temperature, much of it condenses as water droplets. By itself this water poses no health and safety issues, except when other compounds are dissolved within it. However, CO_2 can cause asphyxiation in large concentrations. Neither H_2O or CO_2 are routinely monitored either on board the vehicle during operation or at test stations during servicing and maintenance, not being regulated by law.

CO_2 monitoring has, however, been conducted to measure the fuel economy of engines using the carbon-balance method. This assumes that the mass of carbon in a quantity of fuel is equal to the mass of carbon found in the exhaust when that fuel is combusted [10]. Carbon is found in three emissions within the exhaust: HC, CO, and CO_2. The quantity of HC found in the exhaust represents the uncombusted hydrocarbons, the quantity of CO represents partial combustion, and the quantity

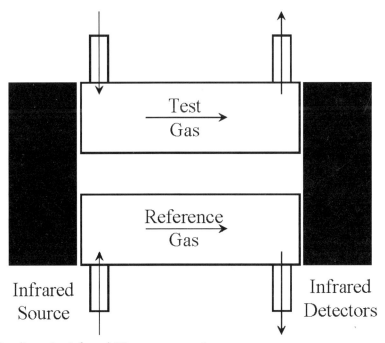

Figure 9.3. Nondispersive infrared CO_2 measurement

of CO_2 represents full combustion. Ideally, full combustion is required to deliver maximum power output from the engine.

There are two main methods for the detection and quantification of CO_2. These are the non-dispersive infrared (NDIR) analysis technique, and gas chromatography. The NDIR technique is a method that can easily be used for the continuous monitoring of CO_2. It compares the amount of infrared light transmitted through a test gas containing the CO_2 with that transmitted through a reference gas. The sensor system consists of an infrared source at one end of the device and a detector at the other end, separated by two gas cells in parallel, as illustrated in figure 9.3. The gas under test passes through one cell, whereas the other contains the reference gas. CO_2 in the test gas absorbs some infrared radiation, and the resultant fall in the received infrared is detected.

It is more difficult to provide a continuous measurement using gas chromatography, since this is a separation method in which the test gas flows over a material column that delays the transport of particular compounds. Different gases are separated due to differences in their partitioning behavior between the mobile gas phase and the stationary phase in the column.

In recent years, CO_2 measurement has been investigated using solid electrolyte sensors, and a lithium phosphorous oxide solid electrolyte sensor seems promising [11]. Sensors using solid electrolytes are commercially available for the environmental monitoring of CO_2 up to about 30000 ppm.

3.2. NITROGEN OXIDES

Nitrogen oxides (NO_x) are emissions that include di-nitrogen oxide (N_2O), nitrous oxide (NO), and nitrogen dioxide (NO_2). NO_x forms when fuels are combusted under high temperature in air. Air contains 78% N_2, which makes it difficult to avoid the formation of these compounds. A typical gasoline engine can produce 2500 ppm of NO_x.

A gasoline engine running at stoichiometry produces mainly NO_x as NO with minimal amounts of NO_2. A diesel engine can produce up to 30% NO_x as NO_2 [12]. Hence, the main pollutant from a diesel engine is NO_x, whereas for the gasoline engines it is CO [13]. Table 9.5 lists the amount of NO_x produced by a gasoline and a diesel engine in a typical journey. Thus, in this context, gasoline engines generally perform better than diesels.

Table 9.5. NO_x production [3]

Gasoline	0.2 g/km	☺
Diesel	0.7 g/km	☹

NO_x emissions cause a number of problems. Acid formation is the first concern because this leads to respiratory problems and forms acid rain, which is detrimental to vegetation. The second largest concern is with ozone formation. NO_x contributes to the formation of ground-level (tropospheric) ozone, again harmful to health and to the environment. In addition, there is concern that NO_x emissions contribute to nutrient overload in soils, a deterioration in water quality, an increase in atmospheric particle density (which lowers visibility), and the production of other toxic chemicals. NO_x and the pollutants formed from NO_x can be transported over long distances. This means that problems associated with NO_x are not confined to areas where NO_x are emitted but can affect larger areas.

NO_x emissions are on the increase. Since 1970, the United States Environmental Protection Agency (EPA) has tracked emissions of the six principal air pollutants: carbon monoxide, lead, nitrogen oxides, particulates, sulfur dioxide, and hydrocarbon compounds. The EPA reported that emissions of all of these pollutants have decreased significantly except for NO_x, which has increased approximately 10 percent over this period [14]. In Europe the NO_x emissions peaked in 1989 at 28 Mt per year and across the European Union of the then fifteen countries, this reached 16 Mt per year [5]. Approximately 50% of this comes from road transportation.

Considerable research is underway for the onboard vehicle monitoring of NO_x [15–23]. One particular type of technology is a development of the zirconia (zirconium oxide) oxygen sensor illustrated in figure 9.4 [16,23]. Here, the exhaust gas diffuses into the O_2 extraction cavity where any oxygen gas is removed by the oxygen pump technique. Two platinum (Pt) electrodes are fitted across the zirconia with one electrode open to the extraction cavity and the other open to air. The platinum electrode on the cavity side catalyses the reduction of oxygen gas to oxygen ions, which are then transported through the zirconia when driven by an external voltage. Once the oxygen ions reach the electrode at the air side (the reference electrode), oxygen gas is formed once more. The exhaust gas, less oxygen, diffuses into the oxygen-free cavity where the sensing of NO_x takes place. A similar electrode arrangement is fitted to this cavity, where NO_x is reduced, so liberating oxygen, which is further reduced to oxygen ions at the Pt electrode. An oxygen pump arrangement or potential difference monitoring arrangement can be used to measure the level of oxygen that has been liberated from the NO_x. Thus, a measurement of the concentration of that NO_x can be achieved.

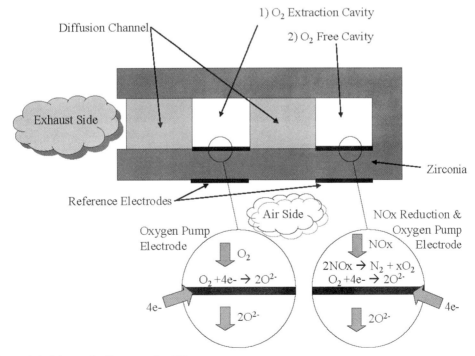

Figure 9.4. Schematic diagram of a NO_x sensor

3.3. CARBON MONOXIDE

In the gasoline engine carbon monoxide (CO) is a major pollutant, which can constitute up to 14% of the volume of the exhaust gases. It is produced because of incomplete combustion, that is, when carbon is not fully oxidized. CO is a colorless, odorless gas, which is extremely toxic. Table 9.6 gives typical figures for the CO produced by gasoline and diesel engines. The latter clearly perform better in this context.

Table 9.6. CO production [2]

Gasoline	1.5 g/km	☹
Diesel	0.6 g/km	☺

CO is extremely toxic because it combines with hemoglobin in the blood to form carboxyhemoglobin. Normally, hemoglobin collects oxygen from air when breathing in, after which blood distributes it to tissues throughout the body. Once the hemoglobin off-loads oxygen, it then collects carbon dioxide, which is then expelled when breathing out. However, the affinity of hemoglobin for CO is 200 times greater than for oxygen, so if CO is present in the air, it is collected in preference to oxygen. Also because of this high affinity for CO, hemoglobin is unable to release it quickly. As a consequence, the tissues in the body are deprived of oxygen and can be damaged. Air containing around 7% of CO can lead to an average person having 50% of their hemoglobin damaged.

Even though CO is extremely toxic, less effort has been put into developing CO sensors than for NO_x sensors. Two major types of CO sensor are under development for automotive applications. Firstly, sensors using Anatase or titanium dioxide (TiO_2) have been investigated [24–26] and have been proposed as a candidate for detecting CO in exhaust gases. Anatase is also sensitive to HC, but by adding CuO its selectivity to CO can be improved. Lanthanum oxide (La_2O_3) can be added as a

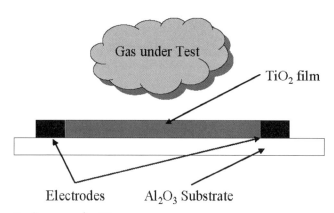

Figure 9.5. Schematic diagram of a CO sensor

structural stabilizer. Secondly, Dutta and others [25] have used stannic (tin) oxide (SnO_2) as the sensing material [13].

Both of these materials are semiconductors whose conductivity increases with an increase in the concentration of CO—and indeed of many other gases. An arrangement of electrodes can be used to measure this conductance change, as shown in figure 9.5.

3.4. HYDROCARBONS

Uncombusted hydrocarbons (HC) in engine exhausts and evaporated fuel have led to raised levels in the atmosphere. In Europe the amount of nonmethane hydrocarbons emitted into the atmosphere peaked in 1989 at a rate of 23 million tonnes per year [5]. HCs are an environmental concern because they are precursors to tropospheric ozone and smog. In addition, HCs themselves can cause health problems, and table 9.7 lists the amounts produced by typical engines. Diesel engines generally perform better in this context.

Table 9.7. HC production [2]

Gasoline	0.2 g/km	☹
Diesel	0.1 g/km	☺

Work conducted on HC sensors for automotive applications has involved perovskite, or calcium zirconate ($CaZrO_3$). Perovskite transducers are based on proton conduction through oxygen vacancies in the structure. The conductivity is determined by the oxygen partial pressure in the atmosphere. Unfortunately, reducing gases such as CO also cause problems.

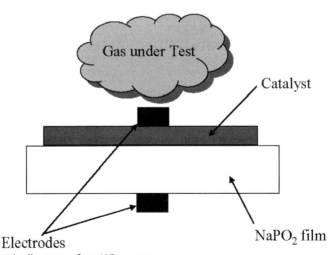

Figure 9.6. Schematic diagram of an HC sensor

Sensors based on proton conduction in alkali orthophosphates have produced a better performance because the conductivity is not dependent on oxygen vacancies and O_2 partial pressures. Cubic sodium orthophosphate has performed well and a typical arrangement is illustrated in figure 9.6.

3.5. PARTICULATES

Carbon particulates are formed when a fuel is incompletely burnt. The larger, heavier hydrocarbons tend to form this type of pollutant. Diesel engines produce nearly 100 times more particulates than gasoline engines, as shown in table 9.8. Carbon particulate matter suspended in the atmosphere has been identified as a health hazard.

Table 9.8. Particulate production [2]

Gasoline	0.01 g/km	☺
Diesel	0.90 g/km	☹

Particulate sizes from natural and human sources can range from nanometers to millimeters. Figure 9.7 illustrates this range with some examples of particulate sources. Large particulates, more than 10 µm diameter, quickly deposit out of the atmosphere, whereas finer particles can remain airborne indefinitely. Particulates classified as PM10 are 10 µm in diameter or smaller. Exhaust emissions from automobiles have diameters up to 25 µm and can be classified as PM25. The PM10 classification is often the more important because of the retention of these particulates in the atmosphere.

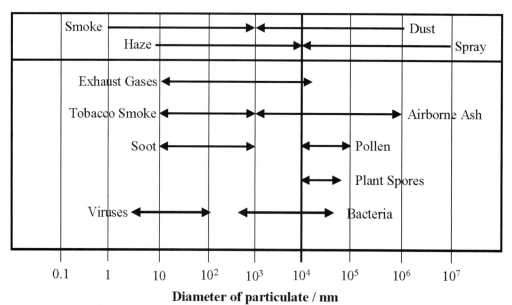

Figure 9.7. Sizes of different particulates [5]

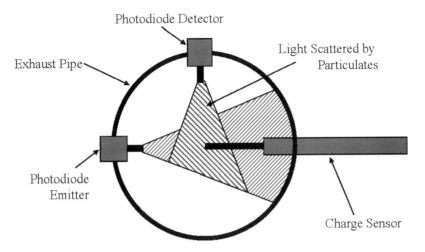

Figure 9.8. Schematic diagram of a particulate sensor

Figure 9.7 also shows the observed effects of particulates. Liquid droplets less than 10 μm produce a "haze", whereas large droplets result in a "spray." Solid particulates of less than 1 μm are known to produce "smoke", whereas particulates larger than this produce "dust."

Particulate detection devices have been developed specifically for automotive applications. A device for monitoring the amount of particulate matter in the exhaust was proposed by Kittelson and others and is shown in figure 9.8 [27]. It involves measuring the electrical charge developed on the sensor as the particulates flow past. Here, either an "image charge" develops or tribo-charging takes place. Figure 9.8 also shows a photodetection system technique used by Kittelson and others to calibrate the charge sensor. However, photo-sensing systems have not been generally successful in automotive applications to date because of the harsh environment.

Bosch and others [28] have investigated a self-regulating particulate sensor based on a sensing plate. As particles adhere to this plate, its resistance changes, giving a measure of particle concentration. Bosch and others propose a method of heating to refresh the plate surface, so that it can repeatedly measure the density of particulates in an exhaust stream

4. EXHAUST GAS OXYGEN SENSORS

Exhaust gas oxygen (EGO) sensors have been the most successful exhaust gas transducers for controlling the performance of the internal combustion engine. In this section the main principles of the exhaust gas oxygen (EGO) sensor are introduced, with particular reference to the zirconia type. This is the most widely used sensor for controlling engine performance and is usually fitted close to the exhaust outlet. Figure 9.9 is a schematic cross section of a typical EGO sensor arrangement using a thimble-shaped sensing component.

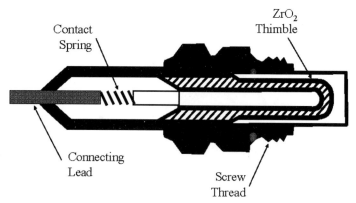

Figure 9.9. Schematic cross section of an EGO sensor

4.1. BASIC PRINCIPLES

An EGO sensor consists of a zirconia (ZrO_2) ceramic solid electrolyte [1,4]. This electrolyte is positioned between two noble metal electrodes, typically platinum, which acts as a catalyst. If the partial pressure of oxygen gas (O_2) is different at the two electrodes, oxygen will flow from the high-pressure side to the low-pressure side as shown in figure 9.10.

The platinum electrode catalyses the reversible reaction of oxygen disassociation and combination as
$$O_2 + 4e^- \leftrightarrow 2O^{2-}$$

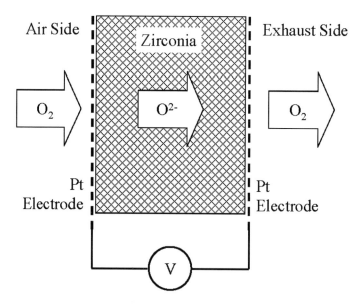

Figure 9.10. Schematic diagram of the flow of O^{2-} through the ZrO_2 ceramic

Oxygen from the "air side" of the sensor forms oxygen cations, which migrate across the zirconia to recombine, forming oxygen gas at the exhaust side. A potential difference is, therefore, developed between the two platinum electrodes given by the Nernst equation (9.2)

$$V = \frac{kT}{4e} \ln \frac{P(O_{2\,air})}{P(O_{2\,exhaust})}, \quad (9.2)$$

where k is the Boltzmann constant, T is the absolute temperature in Kelvins, e is the charge on an electron, and $P(O_2)$ is the partial pressure of oxygen gas. The 4 is in the denominator because there are two oxygen cations carrying a charge of 2- each.

4.2. THE LAMBDA CONTROL PRINCIPLE

The Lambda control method uses the EGO sensor as a sensitive switch to indicate transition through the stoichiometric level from rich to lean burn. Under the rich burn condition (about $\lambda = 0.8$), the engine takes in a higher ratio of fuel to air. As a consequence, the exhaust gases contain a minimal amount of O_2. Figure 9.11 is a schematic diagram illustrating the behavior of the EGO sensor under this rich-burn condition. The exhaust side of the sensor has a lower partial pressure of O_2 compared with the air side. It is this partial pressure difference that drives the diffusion of oxygen species across the zirconia electrolyte. O_2 at the air side is ionized at the platinum (Pt) electrode, forming O^{2-}, which diffuses across the zirconia electrolyte to recombine at the Pt electrode at the exhaust side. Hence, a potential difference is produced between the two Pt electrodes, which can be detected using an appropriate voltmeter.

In the lean-burn condition (about $\lambda = 1.2$), the engine takes in a lower ratio of fuel to air. Consequently, the exhaust gases contain a maximum amount of O_2. The exhaust side of the sensor has a slightly lower partial pressure of O_2 compared to the air side. Hence, a minimal quantity of oxygen

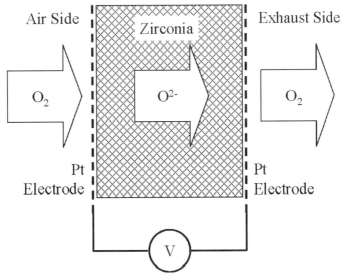

Figure 9.11. Schematic diagram of increased flow of O^{2-} through ZrO_2 ceramic

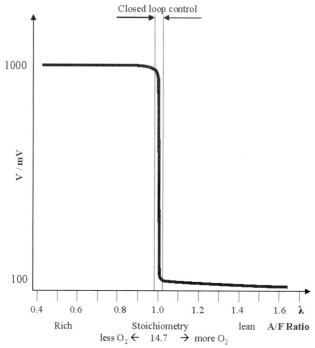

Figure 9.12. The Lambda characteristic

species diffuse across the zirconia electrolyte and a smaller potential difference appears between the two Pt electrodes compared with the rich-burn condition.

Figure 9.12 is a plot of the potential difference across the EGO sensor against the Lambda value. At rich-burn, the potential difference is around a volt, with a sudden drop at the stoichiometric level to hundredths of a volt.

The EGO sensor works most effectively over a range of temperatures from 350 to 800 °C. Below this range the sensor impedance becomes very high, resulting in difficulties in maintaining the voltage output. Above 300 °C the response time becomes quicker, and, as a consequence, it is usual to heat EGO sensors to at least 350 °C.

The Lambda closed-loop control method is used in engine management systems to ensure that an engine combusts fuels at around the stoichiometric level. It monitors the potential difference signal from the EGO sensor and regulates the fuel-air ratio at the intake of the engine accordingly.

4.3. LEAN-BURN CONTROL

This method uses the EGO sensor as an oxygen pump under lean-burn conditions. Figure 9.13 is a schematic illustrating the behavior of the EGO sensor as an oxygen pump. Under lean-burn conditions (about $\lambda = 1.6$), the exhaust side of the sensor has a lower partial pressure of O_2 compared to the air side. The amount of O_2 in the exhaust gas is always lower than on the air side. The leaner the air-fuel ratio, the greater the amount of O_2 found on the exhaust side. If the engine is operated at $\lambda = 1.6$,

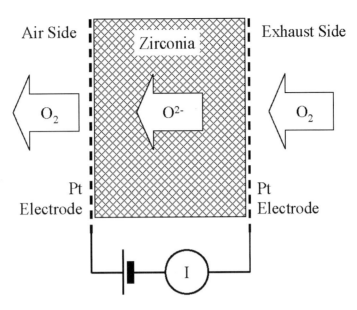

Figure 9.13. O^{2-} pumping

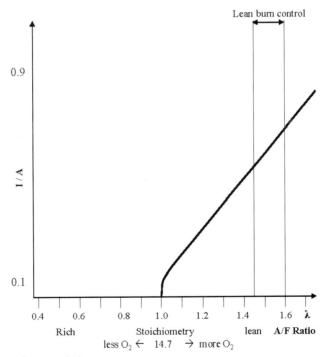

Figure 9.14. Current characteristic

there are significant amounts of O_2 in the exhaust side, which can be pumped to the air side. Figure 9.13 shows an electrical circuit connected to the sensor, which drives electrons to the Pt electrode on the exhaust side, so enabling the formation of O^{2-} ions, and these are pumped to the Pt electrode on the air side. Higher partial pressures of O_2 on the exhaust side lead to higher currents as measured by an appropriate ammeter.

Figure 9.14 shows the current characteristic of this type of sensor. If the engine is operated in the rich-burn condition, the partial pressure of O_2 is insignificant at the exhaust side, so that no current is detected by the ammeter. However, if the engine is to be operated to give best emissions, a lean-burn control method is often used with Lambda values between 1.4 and 1.6.

4.4. ALTERNATIVE OXYGEN SENSOR TECHNOLOGIES

An alternative to the zirconia (ZrO_2) EGO sensor is the titania (TiO_2) EGO sensor, which relies on changes in electrical resistance in the TiO_2 [4], this being a function of temperature and O_2 partial pressure in the exhaust gas. Hence, the sensor does not need to have an air side. TiO_2 is a semiconductor where deficiencies in oxygen ions in the titania crystal determine its conductivity. Higher oxygen partial pressures lead to fewer deficiencies and hence to lower conductivity.

Figure 9.15 is a schematic diagram of a TiO_2 EGO sensor, which has a similar voltage characteristic to the ZrO_2 sensor. The TiO_2 sensing film is connected to a voltage divider circuit with a compensating resistor selected to the midpoint on a logarithmic scale between rich and lean-burn values. The voltage drop across the compensating resistor gives values between almost zero for lean-burn and close to full value under rich-burn conditions. A heating grid is fitted to this design to maintain a constant temperature during operation.

The advantages of the TiO_2 sensor are simplicity of fabrication, compactness, and hence low cost. Reliability questions have been raised, however, so this variant has not displaced zirconia sensors.

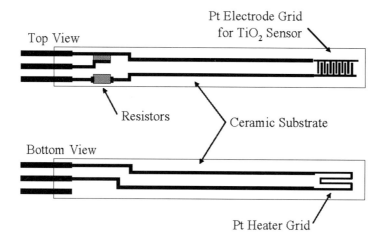

Figure 9.15. Schematic of a TiO_2 sensor

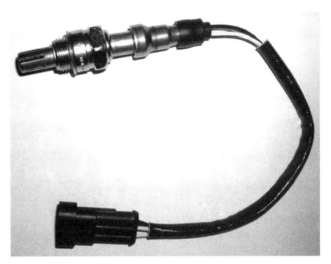

Figure 9.16. Photograph of a EGO sensor

Other materials have been investigated as EGO sensors, which exploit the change in resistance due to oxidation/reduction of the sensor film when exposed to oxygen. These include cobalt, niobium and tin oxides, strontium, lanthanum, and thin-film sensors.

5. EGO SENSOR APPLICATIONS

A typical EGO sensor is illustrated in figure 9.16 with its cable and connector. This particular type is a heated EGO, or HEGO, and is fitted with a heating element to ensure that the sensing component is maintained at its operating temperature for optimum performance. This provides better accuracy, particularly under lean-burn operation.

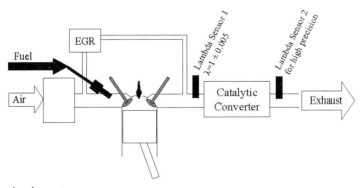

Figure 9.17. Engine layout

5.1. ENGINE CONCEPT DESIGN

The modern gasoline engine is fitted with an engine management system that makes multiple use of the EGO sensor for a number of reasons, including optimization of the torque output or fuel consumption and the minimization of pollutant emissions. It involves a closed-loop system that feeds back information from the exhaust gases to change the air-fuel ratio to achieve the desired engine performance.

Figure 9.17 is a schematic diagram of an engine layout. Air and fuel are mixed at a particular ratio and enter the cylinder. The fuel is combusted and exits the cylinder, so passing Lambda sensor 1, which provides information on the quantity of O_2 in the exhaust gas and hence the air-fuel ratio or the λ value. This information is used to inject more or less fuel into the cylinder at the next combustion cycle, as appropriate. A three-way catalytic converter removes the three main pollutants, NO_x, CO, and HC. Lambda sensor 2 is an optional sensor for closer control of the λ value. It also can indicate deterioration of the catalytic converter and of the upstream Lambda sensor.

An additional option is to incorporate exhaust gas recycling (EGR) where a proportion of the exhaust gas is fed back into the air intake. An air intake of up to 30% of recycled exhaust gas can more than halve the output of NO_x pollutants.

5.2. LAMBDA CLOSED-LOOP CONTROL

Closed-loop control at $\lambda = 1$ is one of the most typical approaches. It ensures that the engine operates at $\lambda = 1 \pm 0.005$ by oscillating between a λ of 0.995 and 1.005. The electronic control unit (ECU) acquires data from the EGO sensor and sends instructions to the fuel system to inject more or less fuel into the cylinder as required. This method is illustrated in figure 9.18, which shows the engine and ECU schematic along with a graph of the sensor input to the ECU and the output signal to the fuel injector.

If the EGO sensor indicates that the fuel mixture is rich ($\lambda = 0.995$), then the ECU will instruct the fuel injection system to incrementally decrease the amount of fuel injected, so making the mixture lean. This is repeated until the EGO sensor indicates that the mixture has indeed become lean ($\lambda = 1.005$). The ECU then instructs the injection system to incrementally increase the amount of fuel

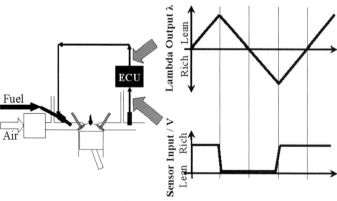

Figure 9.18. Electronic control unit data flow

injected until the sensor indicates that it has become rich. This cycling continues throughout the running of the engine.

5.3. EVEN TIGHTER CONTROL

Since the late 1970s vehicles have been equipped with electronic systems to control emissions and performance. This eventually became standardized, with California's Air Resources Board (CARB) and the EPA (Environmental Protection Agency) requiring vehicle manufacturers to include a self-diagnostic program capable of identifying any emission-related fault in the onboard computer system. This became known as OBD I. Later, OBD II developed this further and was included with every car sold in the United States after 1996. As a result, engine management systems and the use of EGO sensors became more sophisticated.

The EGO sensor works most effectively at normal exhaust gas temperatures, which range from 350 to 800 °C. However, below 350 °C, during warm-up or low engine load, for example, the sensor impedance is very high, giving an inaccurate measurement. The response time of the sensor is also poor at lower temperatures. These problems were overcome by adding a heating element, resulting in the heated EGO (HEGO) [4]. In the thimble design, the heating element is mounted in the interior of the sensor thimble.

The EGO sensor designed to operate for Lambda control around $\lambda = 1$ can be used in lean-burn conditions with small adaptations. At a $\lambda > 1.05$ the voltage variation with λ is much smaller than at

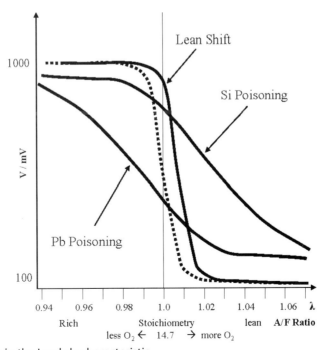

Figure 9.19. Shifts in the Lambda characteristic

around λ = 1, but this can be improved with modifications to the sensor's electrodes to enhance its catalytic activity. A heating element is added and a shield reduces gas flow over the sensor component, so ensuring a constant optimum temperature. EGO sensors adapted for lean-burn operation are called lean-burn EGO or LEGO sensors [4,9].

5.4. PERFORMANCE OF SENSORS

The Lambda sensor characteristic can be influenced by a number of factors. The porous protective layer around the electrode causes the λ curve to shift toward the lean region [1], shown in figure 9.19 as "lean shift." Other lean shifts can be caused by Si poisoning from silicone oils, and also by gasoline, oil, and ash residues forming on the sensor. These tend to plug the pores in the protective layer of the sensor.

"Rich shift" can occur with lead (Pb) poisoning of the sensor, causing deactivation of the electrode. This prevents the formation of O^{2-} ions even if there is O_2 present. Also, fracture in the porous layer of the sensor shifts the λ curve to the rich region, because the response time is reduced.

6. FUTURE PERSPECTIVES

6.1. Emerging Sensing Technologies

Thimble-type EGO sensors are the most commonly used devices, but thick-film and thin-film (or planar) technologies are being developed and are expected to proliferate in the future. Figure 9.20a illustrates a typical thimble sensor configuration, which requires an air reference. By comparison, a

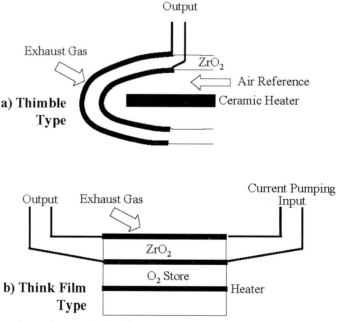

Figure 9.20. Comparison of sensor technology

compact thick-film type of oxygen sensor [29] does not require an air reference, as seen in figure 9.20b. This approach is illustrated by planar oxygen sensors currently under development [18].

Advantages of planar sensors compared with the thimble type include the following:

- High accuracy measurement
- Fast warm-up (light-off)
- Very high temperature operation (1000 °C)
- Low power consumption
- Compact configuration
- No necessity for air reference

6.2. EMERGING VEHICLE TECHNOLOGIES

In all developed countries emission control legislation is in place to limit pollutants emitted by motor vehicles. Eventually, all IC-engined vehicles will be fitted with EGO sensors for engine management and pollution control. In the longer term other gases are expected to be monitored, making for tighter control leading to optimum fuel economy and minimal emissions. With the emergence of new alternative fuels and engines, these sensors will have to be adapted to operate under markedly different conditions.

Gas sensors are already widely employed in engine management. However, future new vehicle technologies are expected to produce new applications for such devices. Climate control systems for the passenger cabin are common, and with the advent of CO and NO_x sensors it is possible to control air quality within vehicles, so protecting the passengers from external pollution. When driving directly behind another vehicle, the concentration of CO typically ranges from 30 to 100 ppm and that for NO_x from 2 to 10 ppm [13]. When the pollutant level is high, the air intake to the cabin can be minimized or filtered. The use of filters only when necessary increases their lifetime compared with continuous use.

One such auto ventilation control system (AVCS) has been reported by Ito and others [20] and uses a sensor for controlling the cabin environment. The sensor monitors two types of gases, a reducing gas such as CO and an oxidizing gas such as NO_2. Ito and others report that driving through a tunnel increases the concentration of CO in the air from about 1 ppm to 15 ppm and NO_2 from 0.1 ppm to 0.4 ppm.

7. SUMMARY

Gas composition sensors for automotive applications are mainly employed in engine management and emission control. Gasoline and diesel engines are the most common power plants in road vehicles, and combustion of the relevant fuels produces water vapor plus carbon dioxide and other unwanted emissions. The normalized air-fuel ratio Lambda (λ) is a parameter that has been developed to monitor the combustion process. At the stoichiometric point ($\lambda = 1$), the ideal air-fuel ratio of the mixture entering the engine, full combustion takes place where all fuel and oxygen are consumed. In reality the ideal condition does not take place and some fuel and oxygen comes through to the exhaust. However, the Lambda parameter is a powerful tool in monitoring the combustion process for identifying the stoichiometric point, and rich- and lean-burn conditions. Near to the stoichiometric point, maximum torque is developed by the engine with minimum HC emissions. However, maximum NO_x production also occurs at this point. This may be overcome by operating the engine under lean-burn conditions, which give good fuel economy and low emission for CO and NO_xs.

Due to pollution problems, legislation is now in place in most developed countries to minimize vehicle emissions, and this has led to continued work on sensors for NO_x, CO, HC, and particulates with the objective of achieving tighter control of emissions. Gasoline and diesel engines each have particular problems with different pollutant emissions. Gasoline engines perform better than diesel engines in producing lower emissions of NO_x and particulates, whereas diesel engines are better at producing lower emissions of CO and HC.

The exhaust gas oxygen (EGO) sensor is probably the most successful automotive gas sensor and has been available since the mid-1970s. It monitors the oxygen content of exhaust gases according to the characteristic Lambda curve. Under rich-burn conditions the EGO sensor gives an approximately 1 V output, which switches to around 0 V under lean-burn conditions, so defining the stoichiometric point of combustion. This response is utilized in engine management systems within a closed-loop control system that oscillates the air-fuel ratio between a slightly rich and a slightly lean mixture around this stoichiometric point. The sensor can also be used in the oxygen pump mode to give accurate monitoring under lean-burn condition. The modern motor vehicle utilizes both stoichiometric and lean-burn operation for optimum performance and reduced emissions.

Advances in thick- and thin-film manufacturing techniques have made it possible for EGO sensors to be fabricated in a planar form as opposed to the classical thimble type. This has led to a number of technical advantages, as well as savings in manufacturing costs. Planar sensors are compact, more accurate, and exhibit faster warm-up times and lower power consumptions. They also make possible the integration of several gas sensors into one package.

In addition to improvements in engine management, gas sensors are now employed in climatic control systems to minimize the entry of pollutants into the passenger cabin. Such innovative applications will secure the role of gas sensors in the automotive field for the foreseeable future.

REFERENCES

1. Jurgen, R. K. 1999. *Automotive electronics handbook.* 2nd ed. New York: McGraw-Hill.
2. Jones, T. O. 1992. Convergence—past and future. In *Proceedings of International Congress on Transportation. Electronics (Convergence) Conference P-260,* 1–3. Warrendale, PA: SAE.
3. Bauer, H., ed. 1996. *Automotive handbook.* 4th ed. Stuttgart: Bosch.
4. Westbrook, M. H., and J. D. Turner. 1994. *Automotive sensors.* Bristol: IOPP.
5. Lenz, H. P., and C. Cozzarini. 1999. *Emissions and air quality.* Warrendale, PA: SAE.
6. Keeling, C. D., and T. P. Whorf. 1997. Trends online: A compendium of data on global change (http://cdiac.esd.ornl.gov/) and Carbon Dioxide Information Analysis Center, Oak Ridge National Laboratory (http://cdiac.esd.ornl.gov/ftp/ndp001r7/).
7. Idso, S. B. 1991. *Carbon dioxide and global change: Earth in transition.* Amsterdam: Elsevier.
8. Houghton, J. T. 1995. *Report of the intergovernmental panel on climate change.* IPCC. Cambridge: Cambridge University Press.
9. Katoh, K., S. Iguchi, and H. Okano. 1992. Toyota lean burn engine—recent developments. *Proceedings of 13th Vienna International Motor Symposium,* 249–56.
10. Wayne, W. S., N. N. Clark, R. D. Nine, and D. Elefante. 2004. A comparison of emissions and fuel economy from hybrid-electric and conventional-drive transit buses. *Energy and Fuels* 18: 257–70.

11. Lee, C., S. A. Akbar, and C. O. Park. 2001. Potentiometric CO_2 gas sensor with lithium phosphorous oxynitride electrolyte. *Sensors and Actuators* B80: 234–42.
12. Hilliard, J. C., and R. W. Wheeler. 1979. Nitrogen dioxide in engine exhaust. SAE paper 790691.
13. Ingrisch, K., A. Zeppenfeld, M. Bauer, B. Ziegenbein, H. Holland, and B. Schumann. 1997. Chemical sensors for CO/NOx—detection in automotive climate control systems. In *Sensors and Transducers, Pt. 68,* 175–83. Detroit: SAE.
14. EPA. 2001. National air quality 2001 status and trends. U.S. Environmental Protection Agency. http://www.epa.gov.
15. Römer, E. 2001. Amperometric NOx sensor combustion exhaust gas control. PhD diss., University of Twente, Germany.
16. Kato, N., Y. Hamada, and H. Kurachi. 1997. Performance of thick film NOx sensor on diesel and gasoline engines. In *Sensors and Transducers, Pt. 68,* 149–56. Detroit: SAE.
17. Nakanouchi, Y., H. Kurosawa, M. Hasei, Y. Yan, and A. Kunimoto. 1997. New types of NOx sensors for automobiles. In *Sensors and Transducers, Pt. 68,* 157–64. Detroit: SAE.
18. Neumann, H., G. Hötzel, and G. Lindemann. 1997. Advanced planar oxygen sensors for future emission control strategies. In *Sensors and Transducers, Pt. 68,* 165–73. Detroit: SAE.
19. Schalwig, J., G. Müller, O. Ambacher, and M. Stutzmann. 1997. Group III Nitride-based gas sensing devices. *Physica Status Solidi (A)* 185(1): 39–45.
20. Ito, S., Y. Iwasaki, Y. Kimoto, K. Hirai, Y. Koyama, and T. Matsuoka. 2003. Development of air quality sensor. In *Sensors and Transducers, Pt. 105,* 11–16. 2nd ed. Detroit: SAE.
21. Szabo, N. F., P. K. Dutte, and A. Soliman. 2003. A NOx sensor for feedback control and emissions reduction. In *Sensors and Transducers, Pt. 105,* 31–35. 2nd ed. Detroit: SAE.
22. Lenaerts, S., P. Van de Voorde, K-P. Sandow, and F. deBlauwe. 2003. A reliability potentiometric NOx sensor. In *Sensors and Transducers, Pt. 105,* 37–39. 2nd ed. Detroit: SAE.
23. Hasei, M., T. Ono, Y. Gao, Y. Yan, and A. Kunimoto. 2003. Sensing performance for low NOx in exhausts with NOx sensor based on mixed potential. In *Sensors and Transducers, Pt. 105,* 41–47. 2nd ed. Detroit: SAE.
24. Figueroa, O. L. 2003. Development of a test system for exhaust gas sensors in streams with large temperature fluctuations. MA thesis, Ohio State University.
25. Dutta, P. K., A. Ginwalla, B. Hogg, B. R. Patton, B. Chwieroth, Z. Liang, P. Gouma, M. Mills, and S. Akbar. 1999. Interaction of carbon monoxide with anatase surfaces at high temperatures: Optimisation of carbon monoxide sensors. *Journal of Physical Chemistry* B 103: 4412–22.
26. Gouma, P. I., S. A. Akbar, and M. J. Mills. 1998. Microstructural characterisation of sensors based on electronic ceramic materials. *JOM* 50(11). SAE paper 960692.
27. Kittelson, D., H. Ma, M. Rhodes, and B. Krafthefer. 2004. Paper presented at the 10th Annual Diesel Engine Emission Ructions Conference, University of Minnesota, Minneapolis.
28. Bosch, R. H., and D. Y. Wang. 2003. U.S. pat. 6,634,210.
29. Makino, K., H. Nishio, T. Okawa, and K. Yabuta. 2003. Compact thick film type oxygen sensor. In *Sensors and Transducers Pt. 105,* 49–52. 2nd ed. Detroit: SAE.

CHAPTER 10

LIQUID LEVEL SENSORS

Yingjie Lin
Francisco J. Sanchez

1. INTRODUCTION

Liquid level sensors are widely used on road vehicles to monitor the amount of fluid contained within a reservoir. There are many applications, including fuel, lubricant, hydraulic system, coolant, and electrolyte level monitoring. New applications are also emerging. Revised emissions regulations for diesel engines are making it likely that there will be a need to measure the amount of urea carried on board a vehicle to ensure that the selective catalytic reduction (SCR) systems now being introduced function correctly.

There are many technologies that can be applied to fluid level sensing, each of which has advantages and disadvantages. As for any transducer, fluid level sensors may be classified by their accuracy, resolution, durability, environmental tolerance, and cost. For automotive purposes cost and durability are probably the most important factors.

Liquid level sensing techniques can be divided into two categories: discrete level indicators (level switches), and continuous level transducers (which measure fluid level and hence estimate volume).

2. LEVEL SWITCHES

Discrete liquid level indicators or switches indicate whether a fluid surface is above or below the switching level.

The simple float switch is a variation on the floating-arm transducer described later. Float switches may consist of a microswitch actuated by a hinged float. More complex designs are possible, in which discrete outputs are produced as the liquid reaches different levels.

For conductive fluids a level switch may be created by using two contacts and an impedance measurement circuit. If the fluid level is lower than the contacts, and air is the medium separating the

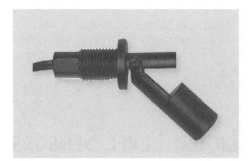

Figure 10.1. Float switch

electrodes, the impedance is high. If the fluid level becomes high enough to wet both electrodes, a low impedance results.

Simple electronics can be used to detect the impedance change. Figure 10.2 shows a typical automotive fluid level switch (note the silver-colored electrodes near the tip of the device).

3. CONTINUOUS LEVEL TRANSDUCERS

A continuous level-sensing transducer makes a measurement (usually in analogue form) of fluid level as long as the liquid surface is within the transducer range.

A number of technologies have been used to design continuous liquid level sensors. The most commonly used technologies are floating-arm sensors, pressure-based sensors, thermal liquid level sensors, ultrasonic level sensors, and capacitive level sensors.

3.1. FLOATING-ARM LEVEL SENSOR

Floating-arm transducers convert liquid level to angular position using a floating arm. A typical example is shown in figure 10.3: the float is the oval component at the bottom of the picture. The most common application for this kind of sensor is the fuel level transducer fitted to almost all vehicles. The angular position of the float arm is normally converted to an electrical signal by a potentiometer. This approach has the advantages of low cost, reasonable reliability, and ruggedness. However, in some applications noncontact angular sensing methods have been used. Examples of such noncontact angular transducers reported in the literature include

- use of a magnetic reed switch array with a moving permanent magnet,
- a rotating magnetic field with field strength sensing, such as a linear Hall effect device,
- a rotating magnetic field with field direction sensing,
- inductive angular position sensing,
- capacitive angular position sensing, or
- optical angular position sensing.

The advantages of floating-arm level sensors are that the technology used is simple and reliable, and the device will work with a multiphase fluid. The disadvantages are that the device

LIQUID LEVEL SENSORS • 261

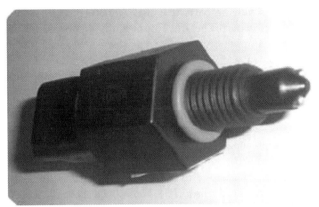

Figure 10.2. Automotive fluid level switch

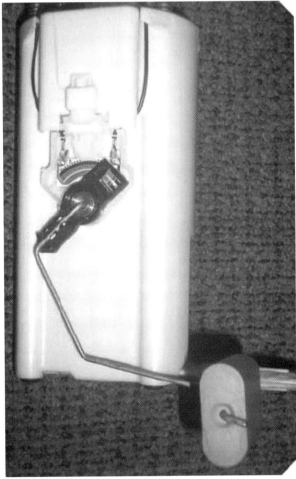

Figure 10.3. Fuel level sensor using resistive strip and floating arm

is essentially mechanical, has moving parts that may wear out, and can suffer from unwanted mechanical effects such as resonance. The arm length limits the measurement range. Some form of electronic or mechanical damping or averaging may also be needed if incorrect readings due to liquid sloshing are to be avoided.

3.2. PRESSURE-BASED LIQUID LEVEL TRANSDUCERS

A pressure-based level sensor converts the pressure reading at a point below the surface of a liquid to liquid level (or depth). A full application of this technique needs three pressure sensors to provide accurate level information. Figure 10.4 shows the basic layout. The first pressure sensor (*P1*) is placed at the bottom of the container. The second transducer (*P2*) is placed a fixed distance above the first sensor, inside the liquid, while the third sensor (*P3*) is positioned above the liquid and measures atmosphere pressure.

The pressure difference between *P1* and *P2*, divided by the vertical distance (*h*) between them, gives the density of the fluid:

$$\text{Density } \rho = \frac{(P2 - P1)}{h}.$$

The difference between *P2* and *P3*, divided by the density of the fluid, is the fluid level (measured from the surface of the fluid to the location of *P2*):

$$\text{Fluid Level} = \frac{(P3 - P2)}{\rho}.$$

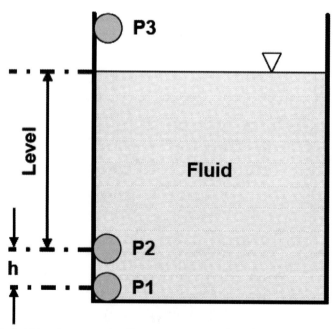

Figure 10.4. Basic fluid level sensor using three pressure sensing elements

For most practical purposes in automotive engineering, *P3* (atmospheric pressure) reading will be constant, and *P3* can be removed. If the fluid density is constant, then *P1* will also be a constant and therefore
$$\text{Fluid level} \propto P2.$$

The main advantage of this form of level sensor is that there are no moving parts. The principal drawbacks are that there is a dead zone (the transducer stops working when the fluid surface falls below *P2*), and that this type of sensor will not work with a multiphase fluid. Pressure-sensor-based liquid level transducers can also be relatively expensive, in part due to the complex signal processing circuits that may be needed.

3.3. THERMAL LIQUID LEVEL SENSORS

In this approach a resistive heater (in the form of a strip) is used as the sensing element. The resistance strip is located down the side of the tank, and the length of the strip defines the measurement range. Figure 10.5 shows the layout of this kind of transducer.

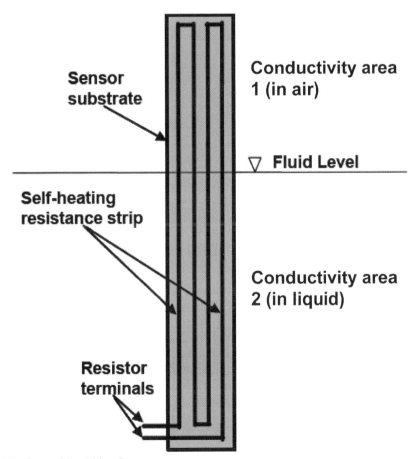

Figure 10.5. Thermal liquid level sensor

To make a measurement the resistance strip is heated using a predefined energy profile, such as constant power or constant current. The differing heat absorption characteristics of air and the measured fluid influence the power required by the heater, and thus the fluid level can be estimated. In fuel the fluid level is the inverse of the self-heating rate: a higher fuel level results in a slower self-heating rate (and hence a reduced power/current requirement), while a lower liquid level results in a faster self-heating rate and increased power.

The advantage of this approach is that the transducer has no moving parts (and hence is reliable). It is also a low-cost solution as resistive heaters and their associated electronics are fairly simple. A resistive heating strip can be applied from the bottom of (for example) a fuel tank, ensuring that there is no dead zone.

The drawbacks are that the accuracy and resolution are generally poorer than for the floating-arm type. Thermal liquid level sensors will not work in a multiphase fluid (unless the thermal properties of both phases are identical!). Finally, unwanted cooling effects can occur due to unforeseen variations in airflow.

3.4. ULTRASONIC LIQUID LEVEL SENSORS

An ultrasonic liquid level sensor operates on the same principle as an ultrasonic distance sensor (a rangefinder). An ultrasonic tone burst is generated, and timing circuits are used to estimate the distance between transmitter and the liquid surface. The excitation pulse(s) may be single step change or a few sinusoidal cycles. The initial excitation pulse triggers a timer. After the initial excitation has stopped, a receiver circuit times the echo response. An echo is caused because acoustic energy is reflected from any acoustic impedance change, such as that created by a liquid/gas interface.

The usual design approach is to define a fixed threshold trigger level. As soon as the echo signal passes this threshold, the timing circuit is stopped. The recorded time is then proportional to the speed of sound within the liquid, which for most practical purposes may be treated as a constant.

In air the speed of sound is around 340 meters/second. In water the speed of sound is around 1400 m/s, and in petroleum the figure is around 1500 m/s.

Because the speed of sound changes slightly with temperature, a reference measurement is normally used to compensate for any thermal changes. This reference is normally provided by using two receiving elements: one that detects the echo returned from the surface of the liquid, and one that detects the echo from an object deliberately placed within the fluid at a known distance from the transmitter. So long as the reference wave-path and the measurement wave-path always travel within the same fluid, the ratio of the two measured times will be proportional to the liquid level and independent of temperature.

Two design approaches are possible and both have been used. Either the ultrasonic level transducer is placed above the fluid (a "face-down" design), so that the acoustic path is in air or it can be placed below the surface of the liquid (a "face-up" design). Figure 10.6 shows the layout used in both cases.

The face-down approach measures the distance from the ultrasound transmitter (in air or vapor) to the surface of the liquid and back. There are some advantages to this: the echo signal is normally stronger, and better accuracy is obtained (since the speed of sound is slower in air/vapor than in a liquid). Packaging requirements are normally simplified if the transmitter is not submerged, and the device will operate with multiphase fluids. The main disadvantage of the face-down approach is that surface waves or sloshing of the liquid will rapidly degrade the measurement accuracy.

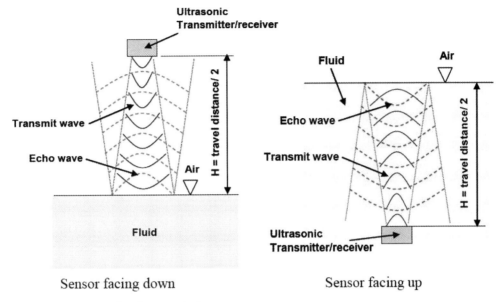

Figure 10.6. Ultrasonic level sensor designs

The face-up design measures the distance from the bottom of the liquid container to the surface of the liquid. A key advantage of this design is that foaming has little impact on the reading. However, the echo signal level is normally lower, and packaging is generally more difficult with a submerged transducer. If a multiphase fluid is present, more than one echo will be generated and accurate measurement may become impossible.

Ultrasonic level sensing has the advantage of no moving parts and hence is very reliable. The resolution is generally adequate for automotive purposes. However, the need for the receiver to be orthogonal to the liquid surface can be a design constraint in some cases. In addition the signal processing requirements, while not particularly complex, can be expensive to implement. Finally, most ultrasound generators are based on the piezoelectric effect and can require high operating voltages. This aspect needs particularly careful handling if ultrasound is to be used close to fuel or other inflammable vapor.

3.5. CAPACITIVE LEVEL SENSORS

A capacitive liquid level sensor makes use of the differing dielectric constants of air and the sensed fluid. Air has a relative dielectric constant of around 1. Most fluids used in road vehicles have a relative dielectric constant of more than 2. For example, fresh engine oil has a relative dielectric constant between 2.0 and 2.1. Vegetable oils (such as olive oil) have relative dielectric constants of 3 or more.

Capacitive liquid level sensors operate by allowing the measured liquid to form a variable-height dielectric between a pair of conductors. Clearly capacitance electrodes that cover the whole measurement range are required, from the bottom to the top of the measured fluid. It is also important that the fluid being measured has homogeneous properties: this kind of sensor will not perform well if a multiphase fluid such as an emulsion or foaming liquid is introduced.

A further limitation is that highly conductive (or ionized) fluids cannot be used with capacitive designs.

The dielectric constant of a fluid is a function of its temperature, its condition (such as contamination level), and age-related parameters (such as the oxidation level). In order to achieve high accuracy, compensation techniques are usually needed. Compensation is often provided by the use of a reference cell to measure the dielectric properties of the fluid (C_{MR}). An initial calibration in dry air is required to establish the base line capacitance (C_B) for both measurement and reference electrode sets.

These base line readings measure a combination of the "dry" electrode capacitance and any parasitic capacitance arising from leads and circuit boards. Assume the parasitic capacitance is constant, then

$$\frac{[C_{ML} - C_B]}{[C_{MR} - C_B]} \propto \text{liquid level,}$$

where

- C_{ML} is the capacitance with the transducer in its measuring position;
- C_B is the "baseline capacitance" (see above);
- C_{MR} is the capacitance of the fluid-filled reference cell.

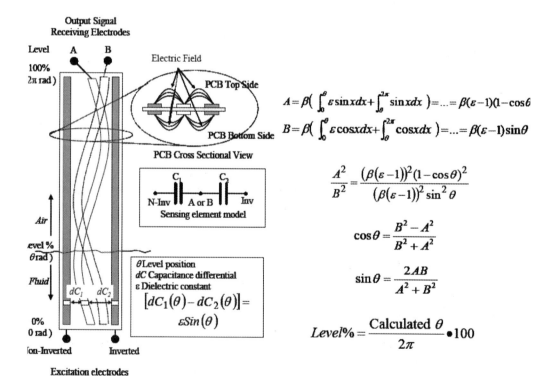

Figure 10.7. Capacitance level sensor

If the parasitic capacitances change with temperature, additional temperature compensation may be required.

A more sophisticated capacitance level sensor design utilizing balanced electrodes has been described [1]. This design uses balanced sine/cosine electrodes that self-compensate for any temperature effects on both parasitic capacitances and the fluid dielectric properties. Figure 10.7 shows this design, further details of which can be found in the reference.

In summary, capacitive liquid level sensors are highly reliable because there are no moving parts or mechanisms. The accuracy and resolution are excellent, and any tendency to display unwanted electromagnetic sensitivity can usually be removed by careful design.

Capacitive sensors cannot be used for conductive (highly ionized) fluids or where more than one phase is present.

3.6. RESISTIVE LEVEL SENSORS

Resistive technology can only be applied to conductive (ionized) liquids, and forms the basis of the liquid level switch described at the beginning of this chapter. Although no automotive applications of this type have been described, in theory it should be possible to construct a resistive analogue of the capacitive transducer described in the previous section. Two parallel conductors would be required, with the current between the conductors being a factor of the extent to which the conductors are immersed in the measured fluid.

If the fluid conductivity is relatively low, then the arrangement used for capacitive level measurement (with reference electrodes) could be utilized for a resistive level sensor. The main difference would be the excitation frequency. For a capacitive sensor the excitation frequency is in the range from several kHz to several MHz. For resistance level sensing, the frequency should be below 100 Hz.

4. SUMMARY AND CONCLUSIONS

The most common form of liquid level sensor used on road vehicles for continuous measurement is still the largely mechanical floating arm device used in fuel tanks. Level switches (many of which also use a float) are used to monitor levels of coolant, hydraulic fluid and (on some vehicles) engine lubricant.

Engine and gearbox oil levels remain a difficult area because of the multiphase nature of the lubricant in its operating condition when large amounts of air can be entrained in the form of bubbles. There is no doubt that a sizeable market would exist if a reliable, low-cost method of monitoring engine oil in particular could be developed.

REFERENCE

1. Lin, Y. 2004. U.S. pat. 6,823,731, B1.

INDEX

Note: *f* indicates figure; *t* indicates table.

aberrant combustion sensors. *See* nonoptimal combustion sensors
absolute angular encoder, 186, 187*f*
absolute sensor, 176–77, 176*f*
absorption dynamometer, 161–63
accelerometers, 19–20, 203–29
    capacitive, 204, 216–19, 217*f*, 218*f*
    definition, 203
    dynamics, 205, 205*f*, 206*f*
    environmental conditions, 222–23
    inertial acceleration switches, 26–27, 27*f*, 28*f*, 219–20, 219*f*
    interconnecting cables, 226–27, 226*f*
    mounting techniques, 223–26, 225*f*, 226*f*
        effects of poor mounting, 224, 224*f*
    optical, 204, 220–21, 220*f*, 221*f*
    piezoelectric, 11, 12*f*, 204, 208–11, 208*f*, 209*f*, 211*f*, 212*f*, 229
    piezoresistive, 204, 211–16, 212*f*, 214*f*
    preamplifiers, 227–28, 227*f*
    selection criteria, 222–29
    sensing principles, 208–21
    theory of operation, 204–8, 206*f*
    various applications, 203–4
acid formation, 240
active infrared (IR) detector, 30–31, 31*f*
active suspension, 18–19
adaptive damping, 19, 19*f*
additives, fuel, 144, 233
adhesion-slip characteristic, 195–97
adhesion-slip curve, 195–96, 196*f*
adiabatic foot, 136
afterglow, 136
airbag sensor, 2, 25–26, 26*f*, 192
air-charge temperature, 122

air conditioning pressure sensor, 73
airflow sensors, 11–14, 107–13
    hot-wire and hot-film flow transducer (MAF sensor), 12, 12*f*, 109–10, 109*f*, 110*f*, 122
    importance of, 107
    moving-vane airflow (VAF), 12, 13*f*, 108–9, 108*f*, 191
    ultrasonic flowmeter, 111, 113, 113*f*
    vortex-shedding flowmeter, 110–11, 112*f*
air-fuel ratio (AFR)
    ion sensing and, 146–47
    lean-burn condition, 247, 248*f*
    lean-burn limit, 120
    luminosity and, 138*f*, 139
    normalized (Lambda), 235–36, 235*f*, 236*f*
    rich-burn condition, 247–48, 247*f*, 248*f*
    stoichiometric, 14, 121, 234*f*, 235, 235*f*
air mass, 122
air temperature sensor, 86–87
alternative fuels, 233
ambient temperature, 122
Analog Devices ADXL50, 217–18, 217*f*
Anatase sensor, 242
anemometer, hot-wire, 109–10, 109*f*
angular acceptance profile, 134
angular encoder, 186, 187*f*
angular movement, 175–76, 175*f*, 176*f*
angular optical sensor, 186–87, 187*f*
angular sensors, 175–76
anodic bonding, 44
antilock braking system (ABS), 20, 193–99
    adhesion-slip characteristic, 195–96, 196*f*, 197*f*
    electronic components, 198, 199*f*
    example calculation, 199
    overview, 193–94
    slip measurement, 197
    variable-reluctance sensor, 198, 198*f*

antispin control, 20
application specific integrated circuit (ASIC), 74–75
argon, 236t
automotive measurands, 190–93
automotive measurement systems, 2, 3f
automotive telematics, 1, 27–29, 32–33
autonomous cruise control (ACC) system, 21, 23
auto ventilation control system (AVCS), 255
axis of sensitivity, 210, 229

ball-and-tube inertia device, 219–20, 219f
barometric pressure, 122
barometric pressure sensor, 72
base metal thermocouple, 92
bellows thermostat, 101–2, 101f
bimetallic temperature sensor, 85, 86f, 95–97, 96f
blind spot, 24
bonding micromachining process, 43–44
bossed diaphragm, 40–41, 47–48, 48f
brake dynamometers
    absorption, 161–63, 163f
    hydraulic, 163–64
brake fluid pressure sensor, 72
brake slip, 195–96, 196f
braking system, 192, 193–94, 194f, 195f
bridge circuits
    for accelerometer, 90, 91f
    for resistance temperature sensor, 90, 91f
    for torque sensor, 168, 169f, 170f
bulk-micromachined capacitive accelerometer, 218–19, 218f
bulk micromachining, 41–42, 41f, 218

cabin environment control, 87, 255
calibration (engine mapping), 116–17
capacitive accelerometer, 204, 216–19
    Analolg Devices ADXL50, 217–18, 217f
    basic structure, 216–17, 217f
    VTI bulk-micromachined, 218–19, 218f
capacitive bridge, 168, 169f
capacitive displacement conversion, 216
capacitive sensors, 50–51
    displacement and position, 50f, 184–85
        characteristics, 185t
        general types, 184f
    driveshaft torque, 166, 167f, 168, 168f, 169f, 170f
    liquid level, 265–67, 266f
    pressure, 10, 11f, 51–52, 51f, 66
    proximity, 25
    tire pressure, 74, 75f

    torque, 166, 167f, 168, 168f, 169f, 170f
carbon-balance method, 239–40
carbon deposition. *See* sooting
carbon dioxide (CO_2)
    atmospheric, 236t
    detection and monitoring, 239–40, 239f
    increasing concentrations, 237
carbon monoxide (CO)
    atmospheric concentration (emission levels), 238t
    detection and monitoring, 242–43, 242f, 242t
    production at different Lambda levels, 235f, 236
    typical exhaust, 237t
carburetor, 107
catalyst, three-way, 14
catalytic converter temperature sensor, 86
ceramic capacitive pressure sensor, 66
cermet inks, 61
Cetane number (CN), 233
charge amplifier, 227–28, 227f
charge coupled device (CCD) camera, 24
Chattock electric wind, 141
chemical combustion sensors, 117
chemi-luminescence, 136, 137, 138
cholesteric liquids, 98–99
circuits
    application specific integrated, 74–75
    bridge (*see* bridge circuits)
    electrical, 140–41, 141f
    four-wire ohmeter, 91f, 92
    hydraulic, 194, 195f
    magnetic, 180–81, 180f
    PN junction sensor, 98, 98f
    thermocouple, 92–93, 92f
climate control systems, 255
closed-loop control
    at Lambda, 252, 252f
    luminosity-based, 139
coherent antistokes raman spectroscopy (CARS), 119
cold junction compensation, 92f, 93, 95f
cold-starting, 121
collision avoidance sensor. *See* crash sensing and sensors
combustion noise, 120, 133
combustion sensors, 16–17, 115–51
    chemical, 117
    design issues, 123–26
    electrical sensors and ion current, 140–47, 148–49t (*see also* electrical combustion sensors)
    for engine input, 121–22, 133, 139, 146–47, 149t
    for engine output, 122–23, 133, 140, 147, 149t

general, 118–19
installation, 124–25
lack of widespread use, 124
nonoptimal or aberrant combustion, 119–21, 132–33, 139, 146
optical sensors and luminosity, 133–40, 148–49t (*see also* optical combustion sensors)
optical signal transmission, 125
packaging requirements, 125
potential applications, 118–23, 147, 148–49t
pressure, 126–33, 148, 148t (*see also* cylinder pressure sensors)
sensor fouling, 125–26
temperature, 117–18
combustion temperature, 119, 138–39
compressor pressure sensor, 73
conductor thick-film ink, 215
cone of reception, 134
continuous liquid level transducers, 260–67
capacitive, 265–67, 266f
floating-arm, 260, 261f, 262
pressure-based, 262–63, 262f
resistive, 267
thermal, 263–64, 263f
ultrasonic, 264–65, 265f
continuously variable transmission (CVT), 18, 73
convoy driving, 29, 33
coolant pressure sensor, 73
coolant temperature sensor, 85–86
crash sensing and sensors, 23, 25–27, 26f, 27f, 219–20, 219f
cross-axis sensitivity, 210, 229
cruise control sensor, 21, 23
cubic sodium orthophosphate sensor, 243, 243f
Curie Point, 209
cycle-to-cycle variation, 119. *See also* nonoptimal combustion sensors
cylinder-head gasket
optical sensor and, 134
piezos and, 129
cylinder pressure, 118
cylinder pressure measurement, 126–27
cylinder-pressure referencing, 130
cylinder pressure sensors, 77–78, 78f, 148–49t
applications, 131–33
cylinder-pressure traces, 131, 131f
design and construction, 127–30, 128f
installation, 130
optical, 129–30
piezo devices, 127–29, 128f

piezoelectric, 16, 56–60, 58f, 59f, 60f, 127, 129
piezoresistive, 46–49, 47f, 49f, 61–62, 127
principles and characteristics, 130–31
purpose of, 126–27
cylinder-to-cylinder variation, 119

damped mass-spring system, 205, 205f
deep reactive ion etch (DRIE), 42
deformation slip, 196, 196f
diaphragm, 38
analytical models, 39–41
bossed, 40–41, 47–48, 48f
bulk-micromachined, 41–42, 41f
deflection, 39, 40f
surface-micromachined, 43
dielectric constant, 265, 266
dielectric thick-film ink, 215
diesel engines
cylinder pressure traces, 131, 131f
ion-current trace, 145
sensor fouling, 126, 140, 143
diesel fuel, 233, 234
diesel injection pressure sensor, 79
diode voltage drop, 97
direct-injection spark-ignition (DISI) engine, 123
discrete liquid level indicator (switch), 259–60, 260f, 261f
displacement and position sensors, 175–201
antilock braking system (ABS), 193–99
capacitive, 50f, 184–85, 184f, 185t
classification methods, 175–77, 176f
future perspectives, 200
hall effect, 189–90, 190t
inductive, 179–84, 181f, 184t
measurands in automotive application, 190–93
based on displacement or position, 191
engine management system, 191–92
parking assist system, 193
safety system, 192–93
vehicle control system, 192
miscellaneous sensors, 190
optical, 185–87, 187t
potentiometric, 177–79, 179t
radar-based, 187–88, 188f, 188t
sensor technologies, 177–90
switch type, 190, 190t
ultrasonic, 188–89, 189t
distance measurement, 24, 25
DPharp resonant pressure sensor, 55
drive-by-wire system, 28–29, 32

driver support. *See* integrated driver support
driver visibility, 24
Druck resonant pressure sensor, 55, 55*f*
dry etching, 42
ds/dt-type measurand, 177
ds-type measurand, 177
duct sensor, 87

electrical combustion sensors, 16–17, 140–47, 148, 148*t*
    applications, 145–47
    design and construction, 140–42, 141*f*
    principles and characteristics, 143–45
electronic control unit (ECU). *See* sensors
emissions, 123
    carbon dioxide, 239–40, 239*f*
    carbon monoxide, 242–43, 242*f*, 242*t*
    characteristic wavelengths, 136, 136*t*
    detection of, 14–16, 238–45
    with different Lambda, 234*f*, 235–36, 235*f*
    hydrocarbons, 243, 243*f*, 243*t*
    nitrogen oxides, 240–41, 240*t*, 241*f*
    particulates, 244–45, 244*f*, 244*t*, 245*f*
    typical motor vehicle pollutants, 236, 237*t*
    water vapor, 239
end-of-travel detection, 191
energy-assisted braking system, 194
engine classification by combustion, 232*t*
engine control module (ECM), 115
    temperature sensors and, 85–86
engine cooling, 100–102
engine input combustion sensors, 121–22, 149*t*
    ion-current, 146–47
    optical, 139
    pressure type, 133
engine mapping, 116–17
engine oil/fuel pressure, 73
engine output (emissions) sensors, 122–23, 149*t*
    ion-current, 147
    optical, 140
    pressure type, 133
    See also emissions
engine output (torque). *See* torque; torque sensors
engine/powertrain management system, 2, 4*f*, 191–92
    temperature sensing applications, 85–86
engine speed sensor, 17, 191
environment of vehicle. *See* vehicle environment sensors
equal time-wise phase shift, 207
etching, 41–42
ethanol, 233

evaporated thin-film pressure sensors, 62
exhaust. *See* emissions
exhaust gas oxygen (EGO) sensors, 14, 15*f*, 245–50, 246*f*
    applications, 251–54
    basic principles, 245–47, 246*f*
    heated EGO (HEGO), 117, 251, 251*f*, 253
    Lambda control principle, 247–48, 247*f*, 248*f*
    lean-burn control, 248, 249*f*, 250
    lean-burn EGO (HEGO), 253
    as oxygen pump, 248, 249*f*
    thick-film (planar), 254–55, 254*f*
    thimble-type, 254, 254*f*
    titania oxygen, 242, 250, 250*f*
    typical, 251, 251*f*
    universal (UEGO), 117
    zirconia oxygen (*see* zirconia oxygen sensor)
    See also gas composition sensors
exhaust gas recirculation (recycling; EGR), 121–22, 252
    effect on ion sensing, 147
    pressure sensors, 72–73
exhaust gas temperature sensor, 86
exhaust oxygen, 123
external combustion, 232, 232*t*
external sensors, vehicle based, 21–25, 22*t*

flame propagation, 119, 139, 145–46
flexural resonators, 57–60, 58*f*, 59*f*, 60*f*
floating-arm liquid level sensor, 260, 261*f*, 262
float switch, 259, 260*f*
fluid level switch, 259–60, 260*f*, 261*f*
foil gauge, 67
force balance, 68
force-balancing, 216
four-wire ohmeter, 91*f*, 92
Froude dynamometer, 163–64
fuel(s)
    additives, 233
        effects on ion sensors, 144
    alternative, 233
    combustion equations, 234
    composition, 122
    diesel, 233
    gasoline, 232–33
    injection scheduling, 121
    pressure sensing, 73
fuel control, 11
fuel economy, 123
fuel injector, 128, 128*f*
fuel pump cutoff switch, 26–27

fusion bonding. *See* silicon fusion bonding

gas chromatography CO2 detection, 240
gas composition sensors, 231–56
    cubic sodium orthophosphate, 243, 243*f*
    emerging technologies, 254–55
    emissions legislation, 238
    exhaust emissions and detection devices, 238–45 (*See also under* emissions)
    exhaust gas oxygen (EGO) (*See* exhaust gas oxygen [EGO] sensors)
    perovskite, 243
    power production and pollution problems, 231–38
    solid electrolyte sensors, 240
    stannic oxide, 242
    titania oxygen, 242, 250, 250*f*
    zirconia oxygen, 241, 241*f*, 245–50, 246*f*
gases, atmospheric, 236, 236*t*
gas gauge, 175
gasoline engine layout, 251–52, 251*f*
gasoline fuel, 232–33
    combustion equation, 234
    See also fuel(s)
gearbox and torque measurement, 164
glow-plug
    ion sensor and, 142
    optical sensor and, 134
    piezo and, 128, 128*f*
Greenhouse effect, 237
Greenhouse gases, 237, 238*t*

Hall effect, 189–90, 189*f*
Hall effect displacement and position sensors, 189–90, 190*t*
hang up, 121
heated EGO (HEGO) gas composition sensor, 117, 251, 251*f*, 253
heat flux gauge, 102–4, 103*f*
heating, ventilation and air conditioning (HVAC) system and temperature sensing, 86–87
helium, 236*t*
heptane, combustion of, 234
hexadecane, combustion of, 234
high-temperature pressure sensors, 69–71
    optical fiber, 71
    piezoelectric materials, 70–71
    silicon on insulator, 70
    silicon on sapphire, 70
homogeneous-charge compression-ignition (HCCI) engine, 123

    cycle-to-cycle variations, 146
    flame propagation, 146
    ion-current trace, 145
hot-film and hot-wire flow transducer (MAF sensor), 12, 12*f*, 109–10, 109*f*, 110*f*, 122
hydraulic circuits, 194, 195*f*
hydrocarbon(s) (HC)
    atmospheric concentration (emission levels), 238*t*
    detection and monitoring, 243, 243*f*, 243*t*
    production at different Lambda levels, 235*f*, 236
    typical exhaust, 237, 237*t*
hydrogen, 236*t*

ignition control sensor, 3, 10
ignition delay, 116
ignition system sensor, 3, 10
    ignition timing map, 10, 10*f*
    ion-current, 141–42
II distribution layout, 194, 195*f*
image-based system, 24–25
incremental angular encoder, 186
incremental sensor, 176–77, 176*f*
in-cylinder combustion measurement, 16–17
in-cylinder pressure sensor, 126–27
in-cylinder soot deposition, 126
indicated mean effective pressure (IMEP), 147
inductive sensors, 179–84
    characteristics, 184*t*
    general types, 181*f*
    linear displacement transducer (LDT), 182, 182*f*
    linear variable displacement (differential) transformer (LVDT), 66*f*, 67, 182–83, 183*f*
    magnetic circuit, 180–81, 180*f*
    pressure, 67
    variable-reluctance sensor, 183
inertial acceleration switch, 26–27, 27*f*, 28*f*, 219–20, 219*f*
infrared detector, 24–25, 30–31
infrared emission and pyrometry, 100
infrared emissions, 136–37, 137–38
inlet manifold pressure sensor, 3, 10, 71–72, 192
insulator thick-film ink, 215
intake air temperature sensor, 86
Integrated Driver Support (IDS), 21–25, 22*t*
    cruise control and collision avoidance (*see* crash sensing and sensors)
    ultrasound, 24
    vehicle environment, 24–25
intelligent transport system (ITS), 200
intensity-based optical pressure sensor, 63, 63*f*
interdigitated transducer (IDT), 56–57, 56*f*

interferometry for cylinder pressure sensors, 129–30
internal combustion, 232, 232t
ion formation, 143
ion sensing, 16–17, 140
    electrical circuit for, 140–41, 141f
    ion-current trace, 144–45, 144f
    See also electrical combustion sensors

kernel development, 119
knock sensors, 10–11, 12f, 120, 132, 139, 146. See also piezoelectric accelerometer
krypton, 236t

Lambda closed-loop control, 252, 252f
    shifts in characteristic, 253f, 254
Lambda control principle, 247–48, 247f, 248f
Lambda sensor. See exhaust gas oxygen (EGO) sensors
laser-based optical rangefinding (LIDAR) system, 23
laser-induced fluorescence (LIF) combustion temperature measuring, 119
lean-burn condition, 247, 248f
lean-burn control, 248, 249f, 250
lean-burn EGO (LEGO) gas consumption sensor, 14, 16, 253
lean-burn limit, 120, 146
lean-burn technology, 14, 16, 238
lean formal method (LFM), 33
lean mixture, 234f, 235–36, 235f
lean shift, 253f, 254
legislation, emissions, 238
linear displacement transducer (LDT), 182, 182f
linear movement, 175–76, 175f, 176f
linear optical sensor, 185–86, 186f
linear variable displacement (differential) transformer (LVDT), 66f, 67, 182–83, 183f
liquid crystal temperature sensor, 98–99, 99f
liquid level sensors, 191, 259–67
    continuous level transducer, 260–67
    level switch, 259–60, 260f, 261f
load, 3, 122–23
loop detector, 29–30, 30f
Lorentz force, 189–90
Lucas optical torsion-bar sensor, 168–71, 170f
luminosity, 133
    applications, 138–40
    measurement, 136–38, 138f

machined piezoelectric crystal sensors, 56–60
    quartz piezoelectric, 60
    quartz resonant, 57–60, 58f, 59f, 60f

surface acoustic wave (SAW), 56–57, 56f
magnetic circuit, 180–81, 180f
magnetic mounting adapter, 226, 226f
magnetorestriction effect, 171
magnetostrictive torque transducer, 171–72, 172f
manifold pressure sensors, 3, 10, 71–72, 192. See also exhaust gas oxygen (EGO) sensors
masking layer, 41–42
mass airflow (MAF) sensor, 109–10, 122. See also airflow sensors
mass fraction burned (MFB), 118, 145
mass-spring damper, 205, 205f
maximum brake torque (MBT), 122
mechanical movement, 175–76, 176f
metal capacitive pressure sensor, 66
metallic resistive temperature sensor, 87–88, 88f
metal-oxide-semiconductor (MOS) transistor, 68
methanol, 233
methylcyclopentadienyl manganese tricarbonyl (MMT), 144
methyl-naphthalene, combustion of, 234
methyl tert-butyl ether (MTBE), 144
M85 fuel, 233
microelectromechanical systems (MEMS), 38, 200. See also silicon micromachined pressure sensors
microswitch displacement and position sensor, 190, 190t
microwave detectors, 31
misfire, 119–20, 132, 146
moving-vane airflow (VAF) sensor, 12, 13f, 108–9, 108f, 191
muscular-energy braking system, 194

negative temperature coefficient (NTC) thermistor, 88–89
neon, 236t
Newton's second law, 203
nitrogen oxide(s) (NO)
    atmospheric concentration (emission levels), 238t
    detection and monitoring, 117, 240–41, 240t, 241f
    production at different Lambda levels, 235f, 236
    typical exhaust, 237, 237t
nondispersive infrared (NDIR) analysis, 239f, 240
nonoptimal combustion sensors, 119–21, 147–48t
    ion-current, 146
    optical, 139
    pressure type, 132–33
normalized air-fuel ratio, 235–36, 235f, 236f

INDEX • 275

octane, combustion of, 234
offset determination, 130
ohmeter, four-wire, 91f, 92
Ohms law, 180
oil/fuel pressure sensor, 73
onboard diagnostics (OBD I and II), 253
open-loop control, 116–17
optical accelerometer, 204, 220–21, 220f, 221f
optical and optical fiber-based pressure sensor, 62–64, 63f
optical combustion sensors, 17, 133–40, 148–49t
    applications, 138–40
    design and construction, 134–36, 135f
    light acceptance profile, 135f
    principles and characteristics, 136–38, 136t
    probe-tip designs, 135f
optical cylinder pressure sensors, 77, 78f, 129–30
optical displacement and position sensor, 185–87
    angular, 186–87, 187f
    characteristics, 187t
    translational linear, 185–86, 186f
optical fiber for high-temperature sensors, 71
optical torsion-bar sensor, 168–71, 170f
optical vortex-shedding sensor, 111, 112f
oxygen, exhaust, 123
oxygen pumping, 14
ozone (O_3), 237, 237t, 240

packaging of MEMS sensors, 44–45
parking assist system, 24, 193
parking braking system, 24, 193
particulates
    detection and monitoring, 244–45, 244t, 245f
    sizes and observed effects, 244f
    typical exhaust, 237, 237t
passivation layer, 41–42
passive infrared (PIR) sensor, 24–25, 30
pegging, 130
perovskite sensor, 243
petrol fuel, 232–33
phase-based optical pressure sensor, 64
phase distortion, 207
phase shift, 207
photolithography, 216
piezoelectric detectors in vehicle-highway systems, 31–32
piezoelectric accelerometer, 11, 12f, 204, 208–11
    cross-axis sensitivity, 210, 229
    design types, 209, 209f
    frequency response, 209, 210f
    micromachined, 210–11, 211f
    schematic, 208f
    See also knock sensors
piezoelectric effect, 204, 208–9, 211
piezoelectric materials for high-temperature sensors, 70–71
piezo pressure sensors, 127–29, 128f
    piezoelectric, 16, 56–60, 58f, 59f, 60f, 127, 129
    piezoresistive, 46–49, 47f, 49f, 61–62, 127
piezoresistive accelerometer, 204, 211–16, 212f
    resonance frequency, 214
    sensitivity, 214–15
    silicon, 213–15, 214f
    thick-film, 215–16
piezoresistive effect, 204
piezoresistivity, 45–46
piezoresistor, 212–13, 213f
planar exhaust gas oxygen (EGO) sensor, 254–55, 254f
plasma etching, 42
PN junction temperature sensor, 97–98, 97f
Poisson's ratio, 45
pollutants. *See also under* emissions
pollution
    global, 237, 238t
    legislation, 238
    local, 236–37, 237t
polymer optical accelerometer, 221, 221f
polysilicon, 46
polysilicon sensor, 49, 49f
position, displacement, and velocity (PDV) sensor, 177
positive temperature coefficient (PTC) thermistor, 89, 90f
potassium hydroxide (KOH), 41
potentiometric sensors, 177–79, 178f
    characteristics, 179t
    pressure, 68, 69f
    signal output linearity, 178, 179f
powertrain management system. *See* engine/powertrain management system
powertrain sensors, 2–17
    emission control, 14–16
    engine speed and torque measurement, 17
    fuel control, 11–14
    ignition control, 3, 10
    in-cylinder combustion measurement, 16–17
    knock sensing, 10–11
    specifications, 2, 5–9t
preamplifier, 227–28, 227f

precious metal thermocouple, 92
pressure balance, 68
pressure-based liquid level transducer, 262–63, 262f
pressure sensors
    applications, 71–79
    high-temperature, 69–71
    operating principle, 38
    piezoelectric, 16, 56–60, 58f, 59f, 60f, 127, 129
    piezoresistive, 46–49, 47f, 49f, 61–62, 127
    types based on fabrication technology, 37–80
        conventional assembly, 64, 66–68, 66f
        machined piezoelectric crystal, 16, 56–60, 58f, 59f, 60f, 127
        optical and optical fiber-based, 62–64, 63f
        other principles, 68
        screen-printed thick film, 61–62
        silicon micromachined, 38–56 (*See also* silicon micromachined pressure sensors)
        sputtered or evaporated thin film, 62
    See also cylinder pressure sensors
probe tips, 126, 135f
proximity detection, 25, 191

Q factor, 53–54
quartz, 57, 59–60
quartz piezoelectric pressure sensors, 60, 127
quartz resonant pressure sensors, 57–60, 58f, 59f, 60f

radar displacement and position sensor, 187–88, 188f, 188t
rangefinder, 193
rate of heat release (RoHR), 118–19
Rayleigh wave, 172
reluctance, 67
reluctive pressure sensor, 67–68
research octane number (RON), 232–33
reservoir fluids. *See* liquid level sensors
resistance temperature sensor bridge circuits, 90, 91f
resistive liquid level sensor, 267
resistive temperature sensors/resistance temperature detectors (RTDs), 87–92
    bridge circuits, 90, 91f
    metallic, 87–88, 88f
    thermistors, 88–89, 90f
resistor thick-film ink, 215, 216
resonance frequency
    accelerometer, 207–8, 207f
    silicon piezoresistive accelerometer, 214
resonant quartz pressure sensors, 57–60, 58f, 59f, 60f

resonant (silicon-micromachined) pressure sensor, 52–55, 53f, 55f
resonator, quartz, 57–60, 58f, 59f, 60f
rich-burn condition, 247–48, 247f, 248f
rich mixture, 234f, 235–36, 235f
rich shift, 253f, 254
robustness, 116
rollover protection sensors, 27, 28f, 193
rough engine running. *See* nonoptimal combustion sensors

safety system, vehicle-based, 25–27, 192–93
sapphire, 70
screen-printed thick film sensor, 61–62
screen printing, 61, 216
seatbelt sensor, 192
seismometer, 206
sensors
    data flow, 252, 252f
    on-vehicle, 2, 3f
        antilock braking and tracking control, 20
        driver support, 21–25 (*see also* Integrated Driver Support [IDS])
        manufacturing methods, 32
        powertrain, 2–17 (*see also* powertrain sensors)
        safety sensors, 25–27, 192–93
        suspension control, 18–20
        tire and wheel sensing, 21
        transmission control, 17–18
    overview, 1–2
    vehicle-highway systems, 1–2, 27–32 (*see also* vehicle-highway systems)
    See also individual sensors
service braking system, 193
silicon, 45, 46
silicon fusion-bonded pressure sensor, 49, 49f
silicon fusion bonding, 44
silicon micromachined pressure sensors, 38–55
    capacitive, 10, 11f, 51–52, 51f
    diaphragm, 38–41
    micromachining techniques, 41–45, 41f, 43f
    piezoresistive, 46–49, 47f, 49f, 127
    resonant pressure sensor, 52–55, 53f, 55f
silicon on insulator sensor, 70
silicon on sapphire sensor, 70
silicon piezoresistive accelerometer, 213–15, 214f
smog, 236
solid electrolyte sensor, 240
sonar, 188

sooting, sensor, 125–26, 236
    in-cylinder sensor, 126
    ion sensor, 143
    optical sensor, 137, 140
spark plug
    as ion sensor, 140–42, 141*f*
    optical sensor and, 134, 135*f*
    piezo and, 128
    pressure sensor and, 77, 78
spatial position measurement, 64, 65*f*
specific fuel consumption (SFC), 123
speed-density airflow sensing, 107. *See also* inlet manifold pressure sensor
sputtered pressure sensor, 62
stannic oxide sensor, 242
start of combustion (SOC), 188
    ion sensing and, 145
    luminosity and, 138
strain gauge
    physics, 45–46
    pressure sensor, 10, 11*f*, 67
    sensitivity, 45
    torque transducer, 164–65, 165*f*, 166*f*
structural damping, 54
s-type measurand, 177
surface acoustic wave (SAW), 172, 173*f*
surface acoustic wave (SAW) devices
    resonator, 56–57, 56*f*
    tire pressure sensor, 76–77, 76*f*
    torque transducer, 172, 173*f*
surface-micromachined pressure sensor, 49–50, 49*f*
surface micromachining, 42–43, 43*f*
surface wave, 172
suspension control sensor, 18–20, 192
switch type displacement and position sensor, 190, 190*t*

telematic system, 1, 27–29, 32–33
temperature, ambient, 122
temperature sensors, 85–104
    bimetallic temperature, 85, 86*f*, 95–97, 96*f*
    combustion, 117–18
    heat flux gauges, 102–4, 103*f*
    infrared emission and pyrometry, 100
    liquid crystal, 98–99, 99*f*
    PN junction, 97–98, 97*f*
    resistive transducer, 87–92, 88*f*
    thermocouple, 92–95, 92*f*, 117–18
    thermostat, 100–102

thermal liquid level sensor, 263–64, 263*f*
thermistor, 88–89, 90*f*
thermocouples, 92–95, 92*f*, 117–18
    arrangements, 93–94, 95–96*f*
    characteristics, 94*t*
    cold junction compensation, 92*f*, 93, 95*f*
thermodynamic loss angle (TLA), 131
thermoelectric temperature sensor, 87
thermopile, 93, 95*f*
thermostats, 100–102, 101*f*
    bellows, 101–102
    wax element, 102
thick-film exhaust gas oxygen (EGO) sensor, 254–55, 254*f*
thick-film material, 216
thick-film piezoresistive accelerometer, 215–16
thick-film pressure sensor, 61–62
thick-film resistor (TFR) ink, 215, 216
thimble-type exhaust gas oxygen (EGO) sensor, 254, 254*f*
thin-film pressure sensor, 62, 77–78
thin-film thermocouple, 117–18
three-way catalyst, 14
throttle position sensing, 191
tire pressure sensing, 21, 73–77
    application specific integrated circuit (ASIC), 74–75
    battery-powered, 76
    pressure-sensing microsystem, 74, 75*f*
    remote, 75–76
    surface acoustic wave (SAW), 76–77, 76*f*
tire slip, 197
titania oxygen sensor, 242, 250, 250*f*
torque
    definition, 159, 160*f*
    maximum brake (MBT), 122
    measurement, 17
    in rotating shaft, 159–60, 161*f*
    usefulness of measuring, 160–61
torque sensors, 3, 17, 159–73
    mechanical, 161–64, 163*f*
    noncontact magnetic, 171–72, 171*f*
    strain-gauge torque transducer, 164–65, 165*f*, 166*f*
    torsion-bar, 165–71, 170*f*
        capacitive, 166, 167*f*, 168, 168*f*, 169*f*, 170*f*
        with optical deflection measurement, 166, 167*f*
torque-speed curves, 161, 162*f*

trace gases, atmospheric, 236, 236t
traction control sensor, 20. See also antilock braking system (ABS)
traffic flow monitoring, 27–28
transducer. See sensors; individual sensors
transient operation, 121
translational linear optical sensor, 185–86, 186f
transmission control, 17–18
transmission fluid temperature sensor, 86
transverse (cross-axis) sensitivity, 210, 229
triangulation, 64
triboelectric noise, 226
two-color combustion temperature measuring, 119

ultrasonic flowmeter, 111, 113, 113f
ultrasonic sensors
    displacement and position, 188–89, 189t
    liquid level, 264–65, 265f
ultrasound, 24
universal exhaust gas oxygen (UEGO) sensor, 117

vane airflow sensor, 12, 13f, 108–9, 108f, 191
variable-reluctance sensor, 183
variable valve timing (VVT), 122, 133
vehicle control system sensors, 192
vehicle dynamic control (VDC) sensor, 192
vehicle environment
    sensors, 24–25, 87
    under-hood, 37
vehicle-highway systems, 1–2, 27–32, 33
    infrared detector, 30–31

loop detector, 29–30
microwave detector, 31
piezoelectric detector, 31–32
video image processing, 32
vehicle speed detector, 30–32
vehicle telematics, 1, 27–29, 32–33
vibration velocity sensor, 206
video image processing, 32
visibility, 24
vision and imaged-based systems, 24–25
voltage amplifier, 227
vortex-shedding flowmeter, 110–11, 112f
VTI bulk-micromachined capacitive accelerometer, 218–19, 218f

water vapor detection and monitoring, 239
wavelength-based optical pressure sensor, 64, 65f
wax element thermostat, 102
Wheatstone bridge circuit, 90, 91f, 228, 228f
wheel speed sensing, 20, 198, 198f. See also tire pressure sensing
wheel-to-body displacement, 19–20

X distribution layout, 194, 195f
xenon, 236t

zero delay (zero phase shift), 207
zirconia oxygen sensor, 15f, 241, 241f, 245–50, 246f
    basic principles, 245–47, 246f
    Lambda control principle, 247–48, 247f, 248f
    lean-burn control, 248, 249f, 250